Basics of Radiation Protection for Everyday Use
How to achieve ALARA: Working Tips and Guidelines

Editors
Harald Ostensen
Gudrun Ingolfsdottir

Author
Leonie Munro

Artist Line Diagrammes
Merle Conway

Creative Digital Imaging
Fiona Walters

WORLD HEALTH ORGANIZATION

WHO Library Cataloguing-in-Publication Data

Munro, Leonie.
Basics of radiation protection for everyday use : how to achieve ALARA : working tips and guidelines /
Editors: Harald Ostensen, Gudrun Ingolfsdottir ;
author: Leonie Munro.

1.Radiography - standards 2.Radiation protection 3.Radiation dosage 4.Quality control 5.Manuals
I.Ostensen, Harald. II.Ingolfsdottir, Gudrun. III.Title.

ISBN 92 4 159178 1   (NLM classification: WN 665)

© World Health Organization 2004

All rights reserved. Publications of the World Health Organization can be obtained from Marketing and Dissemination, World Health Organization, 20 Avenue Appia, 1211 Geneva 27, Switzerland (tel: +41 22 791 2476; fax: +41 22 791 4857; email: bookorders@who.int). Requests for permission to reproduce or translate WHO publications – whether for sale or for noncommercial distribution – should be addressed to Publications, at the above address (fax: +41 22 791 4806; email: permissions@who.int).

The designations employed and the presentation of the material in this publication do not imply the expression of any opinion whatsoever on the part of the World Health Organization concerning the legal status of any country, territory, city or area or of its authorities, or concerning the delimitation of its frontiers or boundaries. Dotted lines on maps represent approximate border lines for which there may not yet be full agreement.

> The mention of specific companies or of certain manufacturers' products does not imply that they are endorsed or recommended by the World Health Organization in preference to others of a similar nature that are not mentioned. Errors and omissions excepted, the names of proprietary products are distinguished by initial capital letters.

The World Health Organization does not warrant that the information contained in this publication is complete and correct and shall not be liable for any damages incurred as a result of its use.

The named authors alone are responsible for the views expressed in this publication.

Printed in Malta

List of editors, authors and collaborators

Editors
Harald Ostensen, M.D., Coordinator, Team of Diagnostic Imaging and Laborat Technology, WHO, Geneva, Switzerland, co-chairman, The Global Steerin Group for Education and Training in Diagnostic Imaging

Gudrun Ingolfsdottir, coordination and support, Team of Diagnostic Imaging a Laboratory Technology, WHO, Geneva,

Author
Leonie Munro, Assistant Director, Radiography, King Edward VIII Hospital, Durban, South Africa

Artist Line Diagrammes
Merle Conway, Durban, South Africa

Creative Digital Imaging
Fiona Walters, Medical Media Services: Nelson R Mandela School of Medicin University of Natal, Durban, South Africa

Foreword

Modern diagnostic imaging offers a vast spectrum of modalities and techniques, which enables us to study the function and morphology of the human body in details that approaches science fiction.

However, it should also be remembered that thousands of hospitals and institutions worldwide do not have the possibilities to perform the most fundamental imaging procedures, for lack of equipment, malfunction or break down of equipment, or insufficient diagnostic imaging skills.

Therefore, WHO in collaboration with The International Commission for Radiological Education (ICRE) of the International Society of Radiology (ISR) is creating a series of manuals and workbooks developed under the umbrella of the Global Steering Group for Education and Training in Diagnostic Imaging. The main issue is to assist and guide "end-users" responsible for diagnostic imaging, be it radiologists, physicians, radiographers, nurses or others to improve safety and quality of their work.

The full series of manuals and workbooks will primarily cover basic examination techniques and interpretation of radiography and ultrasonography as well as radiation safety aspects and basic quality assurance issues .

The manuals are authored by authorities in the specific fields dealt within each manual, supported by a group of collaborators, that together cover the experience, knowledge and needs, which are specific for different regions of the world.

It is our sincere hope that the manuals and workbooks will prove helpful in the daily routine, facilitating the diagnostic work up and hence the treatment, to the best benefit for the patient

Geneva, Switzerland and Lund, Sweden, May 2004
Harald Ostensen
Holger Pettersson

Table of contents

Chapter 1 1
Introduction 1

Chapter 2 3
Production of X-rays 3
    Construction of anode for heat dissipation 5
    Selection of focus spot 6
    Basic X-ray unit components 7
    Risks and benefit for the use of ionizing radiation 8
    Biological effects of ionizing radiation 8
    Factors to minimize radiation dose to patients and staff 8
    Summary 10

Chapter 3 11
X-ray rooms: design, materials and protection barriers 11
    Warning signs 11
    Occupancy of X-ray rooms: calculation of safety aspects 11
Room size 12
    Protective cubicle 13
    Windows and air conditioning units 14
    Doors and walls 14
Tips to ensure doors are closed during radiographic examinations 15
    Walls: materials and lead equivalence 15
Proportions for barium plaster mix 17
    Ceiling and floors 17
    Ward radiography 17
    Change cubicles 18
    Safety measures in special procedure rooms 18
    Summary 18

Chapter 4 19
Radiation protection devices 19
    Lead rubber aprons 19
    Protective lead rubber gloves 20
    Lead rubber gloves 21
    Thyroid shields 21
    Gonad shields 21
    Summary 22

Chapter 5 23
Beam-restricting devices 23
Use of lead blockers to improve image quality 25
Quality assurance tests of beam-restricting devices 25
Working tips 25
Collimator-beam alignment test 26
Method for checking the collimator-beam alignment 26
Test to check alignment of the centre of the X-ray beam 27
Method 27
Use of compression to reduce thickness of patient 29
Summary 29

Chapter 6 31
Scattered radiation: role of grids 31
Grid design 31
Grid ratio 32
Parallel and focused grids 33
Grid cut-off 34
Focused grid: importance of tube-side 36
Stationary grids 37
Moving grids 37
Scatter clean-up: role of grids 38
Care and maintenance of grids 40
Grid factors 44
Summary 44

Chapter 7 45
Radiographic technique, exposure factors, and quality assurance tests 45
Patient positioning 47
Selection of kV 50
Selection of mAs 51
Exposure manipulation: kV/mAs 61
Determining variable kVp charts 62
Quality assurance tests to minimize unnecessary film fog 67
Safelight tests 67
Processor control: performance monitoring 68
Careful film-handling and film storage 70
Summary 71

- Chapter 8 73
  - Exposure to ionizing radiation during pregnancy 73
    - Patients and medical staff 73
    - Pregnant radiation workers 73
    - Summary 73
- Chapter 9 75
  - Self-evaluation of images: application of ALARA..........................75
  - Practical hints for self-evaluation of image quality 75
    - Suggested answers to questions for Figures 9a to 9h..................83

CHAPTER 1

Introduction

Ionizing radiation sources can cause harm both to humans and to the environment. The most important source of ionizing radiation is that used in medicine for diagnostic and therapeutic purposes. Several international guidelines and regulations have been published addressing this aspect of ionizing radiation. The most important of these publications is the *"International Basic Safety Standards for Protection Against Ionizing Radiation and for the Safety of Radiation Sources"*, Safety Series No 11 published by the International Atomic Energy Agency (IAEA), the World Health Organization (WHO), the International Labour Organisation (ILO) and other international organizations. A second publication of major importance addressing the same issues, is the *"1990 Recommendations of the International Commission on Radiological Protection, Publication 60"* published by the International Commission on Radiological Protection (ICRP). It is strongly recommended that these publications are made available to decision makers and relevant medical and technical staff.

It is important not to forget that the primary aim in radiography and radiology is to produce diagnostic images, which assist to establish a correct diagnosis, and thus be of benefit to the treatment of patients. Therefore, the image quality needs to be sufficiently good for diagnostic considerations, i.e., for pattern recognition.

In theory, optimal image quality allows one to make accurate diagnosis. Taking radiation dose into account, however, and to keep this in line with the ALARA principle (**A**s **L**ow **A**s **R**easonably **A**chievable), a certain "balance" between what is optimal and what is acceptable, may be needed. To repeat or not to repeat "sub-optimal" images depends on the clinical situation and the indications for performing an examination. When evaluating images, a decision to re-expose a patient is usually based on experience and a set procedure. If an image is unacceptable, then the radiation received by the patient was not justified, and certainly of no benefit.

The aim of this book is to provide guidance and tips to improve image quality without subjecting patients to unnecessary ionizing radiation

CHAPTER 2

Production of X-rays

In this chapter, the basics of the production of X-rays and risks and benefits of using ionizing radiation for examination of patients, are covered while focusing on the importance of radiation protection:

- A simple definition and explanation of the production of X-rays

- Examples of risks and benefits in terms of radiation dose to patients

- Characteristics and production of X-rays

Like visible light X-rays are part of the electromagnetic spectrum, but the wavelength is approximately 10.000 times shorter (Fig. 2a).

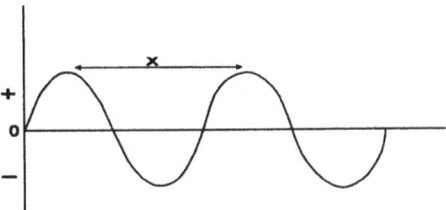

Figure 2a: Drawing of wavelength showing one wave=crest to crest (x).

The short wavelengths of X-rays give them specific properties:

- Ability to penetrate materials such as the human body.

- Ability to induce emission of visible light when hitting certain substances (fluorescence), a phenomenon responsible for reduction of radiation dose needed when using intensifying screens mounted in a light-tight cassette.

- When directed towards a photographic film, the silver halides in the film emulsion are converted to densities that become visible through film development ("processing").

- Ability to induce biologic changes in body cells and tissues, which is the principle behind radiotherapy.

- X-rays are produced in a so-called X-ray tube where the main parts are the cathode (negatively charged) with a filament, and an anode (positively charged).

By means of electrical current measured in milliampere (mA), the filament, which is comprised of coils of wire similar to a light bulb, is heated to glow just as we see in an ordinary light bulb. The difference with the X-ray tube filament is that it does not produce visible light but acts as a source emitting electrons when heated. As the temperature of the filament is raised (this is regulated by the mA settings), more electrons are emitted and the electrical current, or flow, through the X-ray tube increases. The duration of applying this current is expressed in seconds, and in radiography, the product of this current as regulated and measured in *ampere (or, milliampere)* and the time as measured in *seconds,* is called *exposure factor* and abbreviated mAs (milli-ampere-seconds).

The function of the *positive* anode is to attract the *negatively* charged electrons, which are produced by the filament of the cathode. The higher the electrical potential between the cathode and the anode is, the stronger this "attraction" of electrons (i.e., current) will be. The magnitude of this electrical potential, i.e., *difference,* between anode and cathode, is regulated by adjusting the *(kilo) voltage (kV)*.

During application of high voltage across the tube, the electrons impact (collide) with the angled anode and the following occurs:

- Great amounts of heat are produced.

- X-rays of varying wavelengths are produced when the rapidly travelling electrons decelerate (slow down) as they impact with the anode [Figure 2b].

- Ionizing radiation is produced.

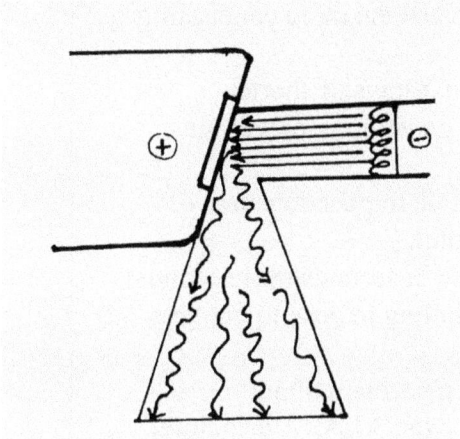

Figure 2b: X-ray tube with electrons flowing from the cathode to the anode. The X-rays are deflected from anode to pass through the tube window.

Construction of anode for heat dissipation

To be able to function properly for long periods (a long "tube life"), the anode must be able to withstand heat. The fast travelling electrons produce heat when they loose energy by impacting with the anode. Dissipation of the heat produced can be achieved by the use of suitable material in a tube.

- The material used in the construction of the anode is usually a block of copper to dissipate the heat.

- Additional material used is a tungsten plate set into the face of the anode in the centre of the tube. Tungsten is used as it has a high melting point allowing the anode to withstand very high temperatures when electrons strike it and X-rays are produced.

> **Tips to reduce dose: Selection of kVp and mAs**
> - The higher the kVp selected the more penetrating the beam = ↓dose.
> - Higher voltage results in X-rays of shorter wavelength and greater penetrating power plus greater intensity.
>
> ♦ Tip: Use highest kV possible to penetrate area of interest (an ALARA principle).
>
> - mAs (tube current x time in seconds/milliseconds) has direct role in contributing to dose to patients. Thus ↑mAs = ↑ dose.
>
> ♦ Tip: Keep mAs as low as possible without compromising image quality (an ALARA principle).

Selection of focus spot

X-ray tubes allow selection of different focus sizes, and the *focal spot* is the area of the anode which is bombarded by electrons from the heated filament. Built into the cathode is a *"focusing cup"*. Its function is to direct the electrons to an area of the tungsten target.

The size of the focal spot (source) has an important effect on the image formed as it is the slightly angled area of the target [Figure 2c] that is struck by the focused 'stream' of electrons. Thus, the smaller the area on the target is, the sharper the image will be.

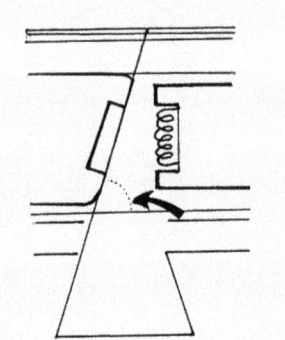

Figure 2c: Angle of anode (arrow).

The use of a small focus depends on the capabilities of the X-ray unit. A large focal spot can withstand more heat than a small one, but some image detail is lost. Focus size for each tube is determined by the manufacturer.

Basic X-ray unit components

A tube requires

- transformer that 'steps-up' the incoming voltage as most exposures include settings from 40k up to at least 150 kVp.

- rectifiers may be needed in countries where the electrical power is supplied as a single-phase alternating current.

- power supplies and controls for the filament, and timers to control duration of exposure.

- protective devices to reduce dose and also to prevent overheating of the tube.

Radiation safety in this context includes shielding of the tube to absorb X-rays emitted in all directions from the focus of the anode. Only the rays that pass through the patient are required for image formation. An X-ray beam consists of different energies of which most contribute to image formation. Low energy radiation, also called 'soft X-rays', do not contribute to image formation but add to the patient dose. Lead lining is used to absorb the majority of X-rays not contributing to image formation. 'Soft X-rays' must be reduced before the beam enters a patient. These low energy rays are absorbed by *inherent* and *additional* filtration. Inherent filtration is the modification of the X-ray beam from the anode. The beam is filtered as it passes through the tungsten deposit on the glass of the tube window, the oil used for cooling of the tube, and the shield aperture. This is specially covered in Chapter 5. Inherent beam filtration in a tube is a prerequisite for radiation protection.

Additional filtration is provided by a shield consisting of aluminium placed outside the tube aperture. The aluminium filter's role is to absorb most of the 'soft X-rays. Total filtration is the sum of the *inherent* filtration and *additional* filtration.

These safety components are mentioned very briefly because their proper functioning is an absolute prerequisite for adequate radiation safety. Regular basic tests should be performed as part of a quality assurance programme to ensure their proper functioning.

Risks and benefit for the use of ionizing radiation

The most important factor to consider is that no radiographic/radiological investigation should be done unless medically justified, i.e., there are sound clinical reasons for the examination, and that the benefit for the patient would outweigh possible radiation risks.

Biological effects of ionizing radiation

A simple explanation of these effects is that when X-rays pass through a patient they cause some biological changes or even damage. Potential damage to the body is of two types, namely *stochastic* and *non-stochastic*, also called deterministic.

- *Stochastic* means something that occurs as a result of the law of chance or probability, and is independent of radiation dose. Stochastic effects, due to exposure to ionizing radiation, can cause cancer, or have influence on gene-material affecting future generations.

- *Non-stochastic (= deterministic)* means that something will always occur, but only when exposure is exceeding a certain *threshold*. The degree of damage (severity) increases the more the threshold value is exceeded.

Factors to minimize radiation dose to patients and staff

Some radiation protection measurements such as filtration of the beam, rectification, and tube shielding are not within a radiographer's control. Others, and these are the main topics dealt with in this book, are within the control of the radiographer/operator, as shown below.

- limitation of field size to area of interest

- use of fast screen-film combinations whenever appropriate

- optimal film processing

- use of automatic exposure timers if available

- use of gonad shields
- selection of grid
- compression of obese patient
- highest practicable kV and lowest mAs
- reduction of number of repeats by careful patient positioning, and use of immobilization devices
- performance of basic quality assurance tests
- no continuous radiation during fluoroscopy
- only required staff allowed into room during radiographic examinations
- all staff should stand behind protective barrier during the exposure
- X-ray units must have adequate shielding
- staff who are required outside the barrier must wear lead-rubber aprons
- field size to be smaller than screen size during fluoroscopy
- staff should stand outside the path of the primary beam, and as far away from it as possible
- lead-rubber flaps to be used on image intensifiers to reduce scatter to staff

Safety measures to reduce dose to patients and staff should also be implemented in operating theatres, and angiography suites. Operators of fluoroscopy units/C-arms, etc., who are not trained in radiation protection measures, should *forced* by national laws to undergo basic training in radiation protection to avoid unnecessary dose to patients, staff, and the environment.

Summary

The production of X-rays for use in medicine could lead to slight chances of damage to living tissue. Provided protection measures are implemented, the risks of potential radiation-induced damage is minimal.

CHAPTER 3

X-ray rooms: design, materials and protection barriers

Most radiographic work is carried out in dedicated rooms (X-ray rooms). Radiation protection measures in these rooms are important because the people in the room, or in the close proximity, could be subject to ionizing radiation, either from the primary beam, or scattered off the patient, or the X-ray table. When considering room design and, all methods should be used to minimize unnecessary exposure to both primary and secondary (i.e. scattered) radiation. All X-ray rooms should allow access only to persons needed for the procedures. Those who work in the room should be protected by means of 'barriers'. These barriers should be thick enough to ensure that radiation dose received does not exceed internationally accepted limits for registered radiation workers.

Warning signs

Warning signs understood by both literate and illiterate persons, must be displayed. These signs must be placed at all entrances to X-ray rooms. In addition, warning lights automatically lit when ionizing radiation is produced, should be installed.

Occupancy of X-ray rooms: calculation of safety aspects

Room size, design, floors, ceiling, doors, and height of windows should be in accordance with national laws and generally accepted international radiation protection recommendations. Based on workload and occupancy factors, radiation safety aspects of X-ray rooms can be calculated by medical physicists [Figure 3a].

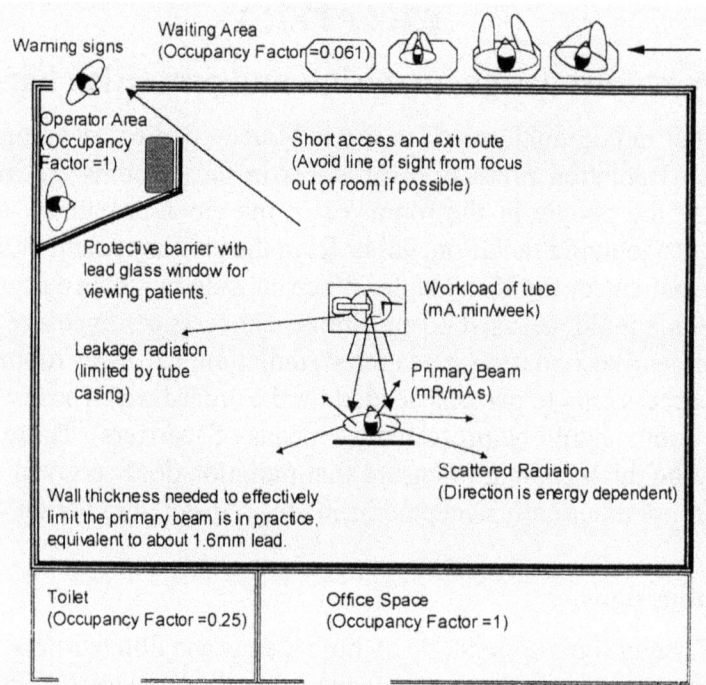

Figure 3a: Example of layout of an X-ray room showing safety aspects that must be considered when designing an X-ray room. Workload and occupancy factors are used by medical physicists to calculate the required thickness of the protective barriers. Protective barriers ensure workers in the room are protected from ionizing radiation. People (arrow) seated in the corridor in close proximity to the room must also be protected (W. Rae acknowledged for the diagram.)

Room size

It is important to note that the size of an X-ray examination room influences on radiation protection. The further one is from the primary beam (X-ray tube), the less radiation dose is received. This is strictly mathematical and based on the inverse square law [Figure 3b], i.e. an increase in distance by two metres reduces the dose by a factor of its "square" (in this case by 2 x 2 = 4).

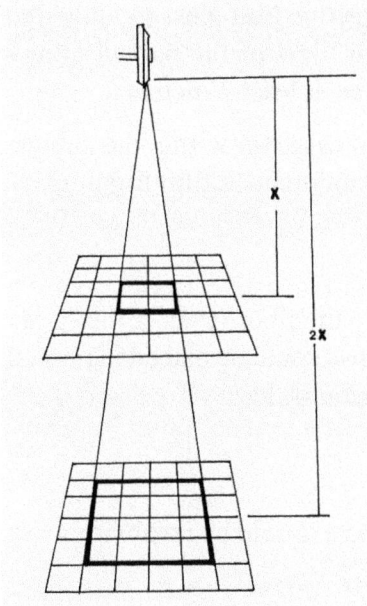

Figure 3b: Diagram showing principle of the inverse square law. As the distance from the source increases, the intensity of the beam decreases proportionally. Beam covers four squares at the distance X but sixteen squares when the distance from source is doubled (2X).

A general purpose X-ray examination room should not be smaller than 16 m^2 to allow safe and adequate installation of equipment. Often, the actual size of existing rooms cannot be altered. When so, proper implementation all radiation protection measurements becomes even more important.

Protective cubicle

The protective cubicle should allow sufficient space to accommodate staff while exposing the films [Figures 3c and d].

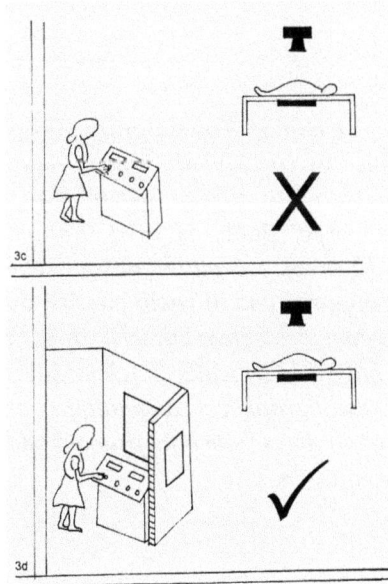

Figures 3c and 3d: Top diagram (3c) does not meet international standards of protection as the radiographer is not protected from primary and scattered ionizing radiation. Bottom diagram (3d) shows worker behind a fixed protective cubicle. Lead glass windows allow clear view of the patient during exposure.

The cubicle should be positioned in such a way that any radiation, be it direct or scattered reaching the radiographer/technologist operating the X-ray unit, is reduced to an absolute minimum. The cubicle should have at least one window with protective lead glass to allow the radiographer / technologist to have a clear view of the patient at any time, and the height of the cubicle should be at least 2 metres.

Storage of exposed and unexposed film cassettes within the cubicle during exposure is recommended to avoid unintended film fogging.

Windows and air conditioning units

These should be at least 2 metres above the floor level. If the X-ray room is above ground level, then the windows could be placed in normal height provided there is no link to passages/corridors.

Doors and walls

The position of doors is important as there should be no obvious risk of exposure to passers-by [Figures 3e and f].

- Sliding access doors give better radiation protection than normal doors; they should overlap each side of the door access/opening by at least 100 mm.

- Doors should be lined with lead sheet of 2 mm thickness.

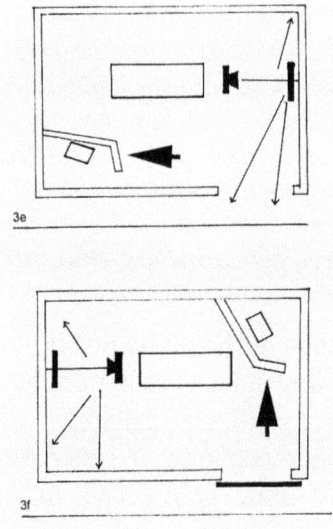

Figures 3e and 3f: Top diagram (3e) is an example of poorly designed room because the door is direct line of X-ray beam when erect Bucky is used. Bottom diagram (3f) is an acceptable design. The door is at opposite end of room out of way of X-ray beam. Protective cubicles in both diagrams have angled walls (thick arrows) to ensure radiographer / technologist is protected at all times from both primary and scattered radiation.

> **Tips to ensure doors are closed during radiographic examinations**
> - Sliding mechanism should be sturdy to hold heavy lead lined doors
> - Doors should be easy to slide open/close
> - Sliding mechanisms should be well maintained and kept clean

Walls: materials and lead equivalence

- Walls should be built with material that absorbs radiation, such as 230 mm baked solid clay bricks

- Lead sheet of 2 mm could be sandwiched between other brick types if needed [Figures 3g,h,i]

- Building blocks with openings require use of lead sheets to prevent radiation passing unhindered through the open areas

- Dry walls (wood/chipboard/plywood, etc.) must have lead linings

- Barium plaster at least 6mm of thickness could also be used to cover the walls. Barium has a relatively high atomic number (56) thereby absorbing some radiation

- Walls should be protected up to 2.2 meters from floor level.

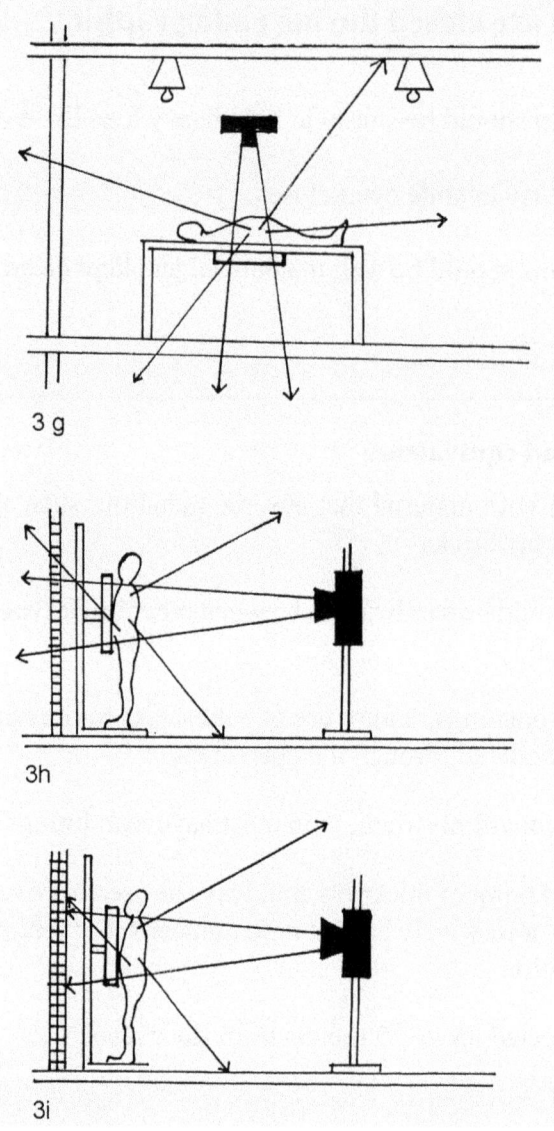

Figures 3g, 3h and 3i: Top diagram (3g) shows primary and scattered radiation passing through ceiling, floor and thin wall as there are no protective barriers in the room. Middle diagram (3h) shows primary radiation passing through a single brick wall which is not an adequate barrier. In the bottom diagram (3i) the room meets international radiation protection requirements as thickness of the walls, floor, and ceiling.

> **Proportions for barium plaster mix**
> - One part coarse barium plaster
> - One part fine barium sulphate
> - One part cement

IMPORTANT REMARK:

The safety of rooms has to be calculated by medical physicists following national laws and regulations. The measurements given above are provided as guidance only.

Ceiling and floors

Ideally, X-ray examination rooms should be on the ground level of a building as this does not require additional protection measures for the floors. If rooms are above ground, the floors should comprise a solid concrete slab of not less than 150 mm thickness. The concrete must be of a high density as recommended by medical physicists (e.g. $2.35 g/cm^3$).

Ceiling slabs must be used in X-ray rooms if the floors above are occupied. Single storey buildings do not require ceiling slabs for radiation protection safety.

Ward radiography

Distance is the most important safety measure to reduce radiation dose to people not undergoing radiographic examination. (see Figure 3b). A minimum of 2 metres from the X-ray tube is usually sufficient provided the X-ray beam is restricted to areas of interest. Beam restriction is discussed in Chapter 5.

Change cubicles

Change cubicles opening into X-ray examination rooms must be lined with 1,5 mm lead sheets, or thicker. Actual thickness would be calculated by a medical physicist. To prevent entrance during radiation exposures, access doors of cubicles should be fitted with locks as a safety measure.

Safety measures in special procedure rooms

The above guidelines for a general X-ray room also apply to special procedure rooms. However, the main aim of this book is to address radiation protection in places with limited resources, where such rooms are rarely available.

Summary

All safety measures available should be implemented to minimize the risks of unnecessary radiation dose to patients, staff and members of the public. These measures include size of room, layout, and materials used. National laws shall be followed. Additionally, there are international guidelines addressing acceptable protective materials used in walls, protective cubicles, floors, and ceilings. Concrete of a specific density and thickness acts as a barrier by absorbing radiation, as do lead linings and lead glass windows. However, the most easily applied important protective measure is increased distance from the source of both primary and secondary scattered radiation.

CHAPTER 4

Radiation protection devices

It is essential that radiation workers be protected when they need to work outside the protective cubicle. There are several essential protective devices, including protective clothing, which should be readily available for use in every X-ray room. These devices are used to protect staff from receiving unnecessary radiation dose from both the primary beam and from scattered radiation. These devices should also be used to shield members of the public from unnecessary dose, for example when a parent holds the arms of a baby during exposure of a chest radiograph.

Lead rubber aprons

As far as reasonably possible, radiation workers such as radiographers, radiological technologists and radiologists should remain in the protected area during exposure. When this is not possible, they should be provided with lead rubber aprons of at least 0.25 mm lead equivalence. If they stand within one meter of the X-ray tube or patient when the unit is operated at tube voltages above 100 kV, they should wear protective lead rubber aprons of at least 0.35 mm lead equivalence. Lead rubber aprons are available as single-sided (protects anterior/front part of body) or double-sided (protects back and front of wearer). If a worker wears a single-sided apron then it is important to always face the source of radiation and not to turn away from the source.

Note that members of the public, who assist during an examination and therefore have to remain inside the examination room during exposure, must be provided with necessary protection devices such as lead rubber aprons and lead rubber gloves.

Care of lead rubber aprons

- To prevent damage to aprons when not in use always hang them up on a sturdy hanger

- Never fold aprons as this could cause cracks in the lead rubber

- Undertake monthly visual inspections of all protective aprons for cracks, splits, rips, tears, etc.

- Aprons suspected to be damaged can be radiographed if in doubt:
  1) Place suspect area of apron on an unexposed loaded cassette and expose to radiation. Using at least 70 kV and 10 – 15 mAs at 100 cm FFD
  2) Process film and inspect for signs of fogging and if noted then withdraw defective apron from use

- Double-sided aprons should be opened fully so that one side at time is checked

- Depending on size and degree of damaged areas, aprons can be repaired. Always re-radiograph a repaired apron to make sure it is suitable for use

- Defective items should not be used.

Protective lead rubber gloves

According to the *ICRP Publication 57*, lead rubber gloves should be at least 0.35 mm lead equivalence. Gloves should be used to protect workers' hands when placed in close proximity or under the primary beam, for example during a barium study. This also applies to any person who is in close proximity to the X-ray beam, such as a parent holding a baby during an X-ray examination [Figure 4a].

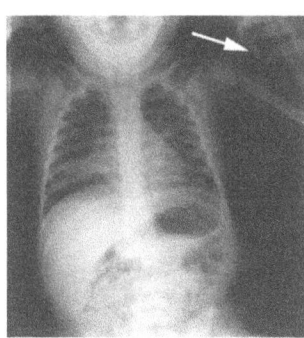

Figure 4a: Arrow shows unprotected hands of helper/parent. This is poor application of radiation protection measures to limit dose to members of the public. It is essential to provide members of the public with lead rubber gloves, or mittens to cover hands within primary beam. Besides the fact that a member of the public received dose to hands, the patient also received unnecessary radiation dose. ALARA was not

implemented as the entire abdomen of the baby was exposed unnecessarily to ionizing radiation. Paediatric chest radiography should always include beam restriction to chest area. This should be done routinely in all paediatric radiography to limit dose to children. This radiograph is a typical example of (i) poor radiographic technique, and (ii) inadequate implementation of radiation protection measures. Also, a black image like this indicates use of too high mAs which contributes unnecessarily to increased patient dose (see Chapter 7).

Lead rubber gloves

- Handle with care to prevent damage

- When not in use, store flat in a safe place within easy reach

- Gloves should be checked monthly for cracks or defective areas. Defective gloves should be withdrawn from use.

Thyroid shields

The thyroid gland is relatively sensitive to ionizing radiation. Therefore, it is recommended to use a radiation protection device whenever possible. There are several types of shields on the market. If not available, a lead rubber apron with a high neckline can be used. Care should be taken when using the shields to ensure they do not get damaged, and they should be stored in a safe place when not in use.

Gonad shields

Whenever possible, gonads should be protected from being exposed to ionizing radiation. When gonads are within the primary beam or within 5 cm of it, some shielding should be used if this can be done without obscuring or excluding information needed for diagnosis.

Gonad shields are of three different types

- Contact shields: these are fairly inexpensive and easy to use as they are made from pieces of lead sheet or lead rubber. Lead gloves can also be used for gonad shielding

- Shadow shields do not come into contact with the patient as they are radio-opaque shields placed between the X-ray tube and the patient

- Shaped contact shields are available for male patients

If a lead rubber apron and lead gloves are beyond repair, parts of these may be cut up and be used as contact gonad shields

Summary

Protective devices/clothing should be used by persons exposed to ionizing radiation related to X-ray examinations of other persons than themselves. Patients should be protected whenever possible provided relevant information is not obscured by the devices.

All protective devices should be inspected on a regular basis to ensure that they are not damaged. The ALARA principle should be applied for every exposure made to patients and this includes use of protective items, whenever possible.

CHAPTER 5

Beam-restricting devices

In accordance with the ALARA principle all possible dose reducing measurements should be employed. Beam-restricting devices play a significant role in dose reduction [Figure 5a].

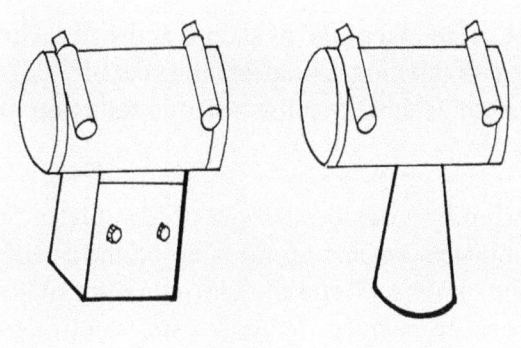

Figure 5a: Line drawing (left) of variable-aperture beam limiting device, i.e. light beam diaphragm (LBD), and (right) removable metal cone.

As scattered radiation adds to dose, all means available should be used to minimize this type of radiation. The larger the area covered by the primary X-ray beam, the greater the amount of scattered radiation produced (Fig. 5b). Apart from contributing to dose, this radiation is in addition to that of the primary beam and adds to film density, which could reduce visualization of important details. In view of this, and to keep scattered radiation as low as possible, the primary beam should be confined to the area of interest for the examination.

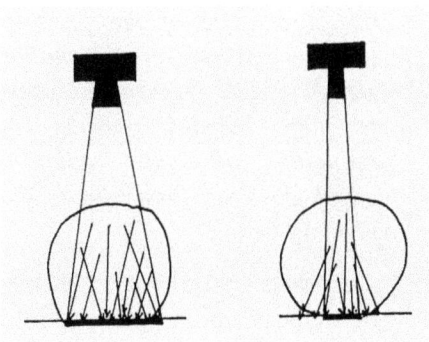

Figure 5b: Diagram (left) of an unrestricted beam; considerable scattered radiation is produced in a thick object. A restricted beam (right diagram) results in production of less scattered radiation; dose to surrounding parts of the object is reduced as no longer in the path of the primary beam.

There are several devices which can be attached to the X-ray tube to restrict the X-ray field size. Beam-restricting devices may be detachable cones or permanent fixtures. Examples of beam-restricting devices include:

- **Aperture diaphragms** are sheets of lead with circular, square, or rectangular openings. These devices are inserted into the X-ray beam near the tube window and are usually used together with a cone or a variable-aperture beam-limiting device.

- **Cones** are metal tubes that come in a range of shapes and sizes. The length of a cone and size of an opening will affect the size of the X-ray field. Cones are usually detachable to allow use of a selection of different sizes and shapes.

- **Variable-aperture** beam-limiting devices consist of lead plates or shutters, which can be adjusted to change the size of the beam, usually by turning indicator knobs. Some modern units are made with a beam-limiting device whose shutters are controlled automatically by sensors adjusting the field size to the size of the cassette (image receptor). These devices contain "cross hairs", a light source, and a mirror to project the light onto the centre point of the patient and to visually indicate outline of X-ray field in accordance with the open shutter size.

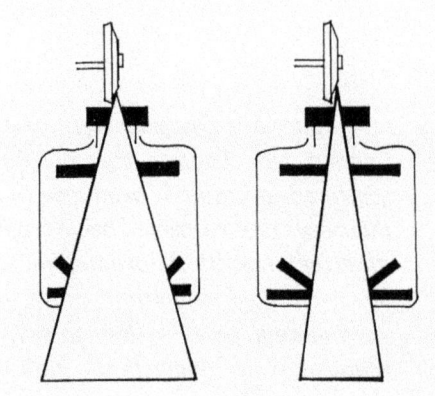

Figure 5c: Example of a variable-aperture, multiple-shutter beam limiting device. Note diagram on the right: size of the primary beam is reduced due to smaller size field. The diaphragm closest to tube window results in reducing off-focus radiation which does not form part of the primary beam.

Apart from the above devices, additional items could

be used to reduce scattered radiation reaching the film. For example, lead blockers can be used to divide a cassette when more than one exposure is made on a single film. Lead blockers reduce unnecessary exposure to both the exposed and unexposed sections of the film thus limiting the chances of film fog and thereby loss of image quality.

Use of lead blockers to improve image quality

- Place a lead blocker on the couch when doing lateral projections of thoracic spine, lumbar spine, sacrum and coccyx spine;

- Position the lead blocker next to the patient's back to absorb radiation that will not pass through the spine. The lead blocker is placed almost in contact with the posterior skin surface of the patient's back;

- Defective lead aprons could be cut into a range of sizes if lead blockers are not available.

Quality assurance tests of beam-restricting devices

It is good radiographic practice to limit the area of radiation so that the resulting image has collimated edges on all four sides of the film. Thus, proper beam-restricting measures (collimation) applied for chest radiographs means minimized or totally excluded radiation to lens of the eyes, as well as other organs not targeted. An example of poor usage of beam-limiting devices is shown in Figure 4a *(page 28)*; the patient received unnecessary radiation dose to the abdomen whilst undergoing chest radiography.

Working tips

- Beam-restricting devices should be checked regularly to ensure they are functioning optimally;

- From time to time during the course of the day, it is strongly recommended to bring the tube to rest on the couch (table top) to check that all four edges of the square light beam diaphragm are in contact with the table/couch top when using a straight vertical beam. Sometimes the tube, due to constant use, moves slightly from its position hence the central ray is not at right angles to the couch.

The same check should be performed with cones when inserted in front of a tube window. Cones sometimes become bent when dropped or bumped and cause irregular and insufficient coning.

Collimator-beam alignment test

- Even if there is no obvious problem, the device should be checked at least every six months to assess proper alignment of the beam-restricting device and primary beam;

- Poor alignment causes sub-optimal positioning of patients as it may be difficult to centre accurately. Thus, the area of interest may not be included on the image, or additional and not wanted areas may be included. In both examples this would lead to increased radiation dose to the patient.

Method for checking the collimator-beam alignment

- Place metal coins or paper clips on a loaded 24 x 30 cm cassette [Figure 5d];

- Set the beam-restricting device (collimator) at 20 x 20 cm and expose the film at 100 cm using 60 kV and 4 – 8 mAs. Process the film and check that the alignment conforms to international performance criteria which allow +/- 2% difference of FFD [Figure 5e];

- Measure the outer edges of the image and the outer edges of the metal coins/paper clips;

- Should the difference exceed the acceptable range of variation, then repair is needed.

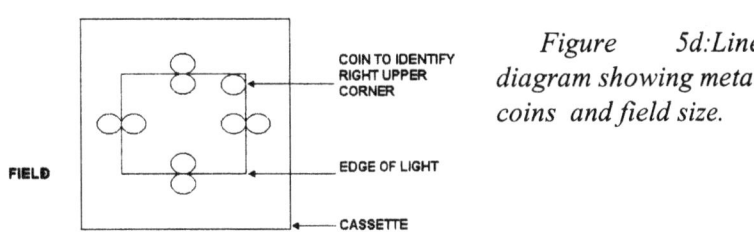

Figure 5d: Line diagram showing metal coins and field size.

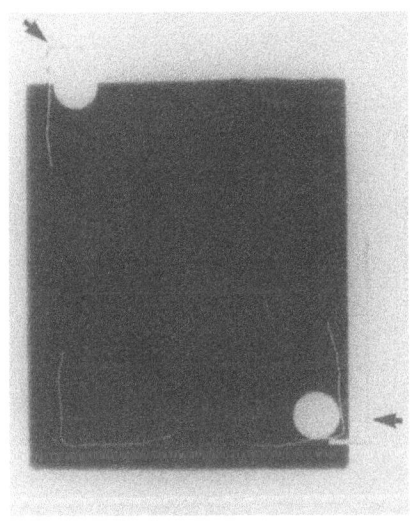

Figure 5e: Arrows indicate light beam edges. Area of film blackening does not overlap light edges. Measurements of both should be taken to determine whether the radiation field size exceeds light field by more than 2% based on the actual FFD.

Test to check alignment of the centre of the X-ray beam

Method

- Place an unexposed loaded cassette in the centre of the bucky tray and centre tube to the cassette;

- Move tube to 100 cm FFD to bucky tray. Reduce the longitudinal collimators to a thin slit (e.g. 0.5 cm). Close the lateral collimators. Expose using 60 kV and 4 – 8 mAs;

- Do not remove the cassette;

- Close the slit collimators and open the lateral ones to collimate laterally to a thin slit (e.g. 0.5 cm). Expose the film again;

- Process the film and inspect the image;

- Bend the film in half and check that the exposed 'cross' is in the centre of film [Figure 5 f];

- This simple test can be used to check alignment of the central ray when performing non-bucky radiography. Remember to increase FFD to measure 100 cm to top of table/cassette;

- A deviation of 1 cm either side of the centre is acceptable, i.e. total of 2 cm;

- Deviations greater than the acceptable range should be corrected to reduce repeat radiographs.

Figure 5f: Arrows indicate middle of film. Cross and film centre are not aligned.

Use of compression to reduce thickness of patient

A further dose reducing measurement is to compress the area of interest, i.e., to reduce thickness of a patient by compression to minimize production of scattered radiation. The thinner the area irradiated is, the less scattered radiation is produced [Figure 5g].

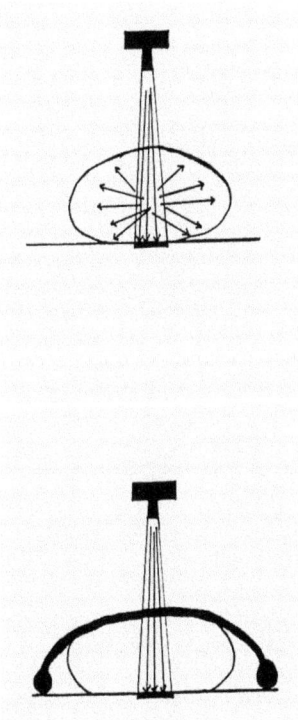

Figure 5g: Example of cross section of abdomen of an obese patient; scattered radiation is in all directions (top). Less scattered radiation produced when overall volume of abdomen is reduced by means of a compression band (bottom).

Summary

Beam–restricting devices play an important role in radiation protection. A device, which is not checked regularly could result in degradation of the image quality and possibly re-exposing a patient. Regular performance of simple quality assurance tests is in accordance with ALARA. The practical implications of the use of beam-restricting devices on exposure factor selection are covered in Chapter 7.

CHAPTER 6

Scattered radiation: role of grids

Scattered radiation should be minimized (controlled) in all radiographic examinations. The use of beam-restricting devices 'control' scattered radiation before the primary beam enters the patient. Reduction of scattered radiation from the patient reaching the image receptor/film is achieved by using a grid. Not only do grids 'clean-up' scattered radiation for overall improvement of the image, but they also reduce the scattered radiation continuing in forward directions after having penetrated the patient. The intensity of the scattered radiation depends on the kV selected; this is discussed in-depth in Chapter 7.

Grid design

A grid consists of alternating strips of lead to absorb scattered radiation, and a spacer material, such as transparent fibre, that does not absorb ionizing radiation. The strips and spacers are encased in a protective firm cover. The cover must be sturdy to ensure the grid strips are aligned correctly to the rays of the primary beam [Figure 6a]. The spacer material is sandwiched between the lead strips and this allows most of the primary rays to pass through onto the film for image formation. The oblique rays of the scattered radiation are absorbed by the lead strips because they are usually at an angle to the lead strips.

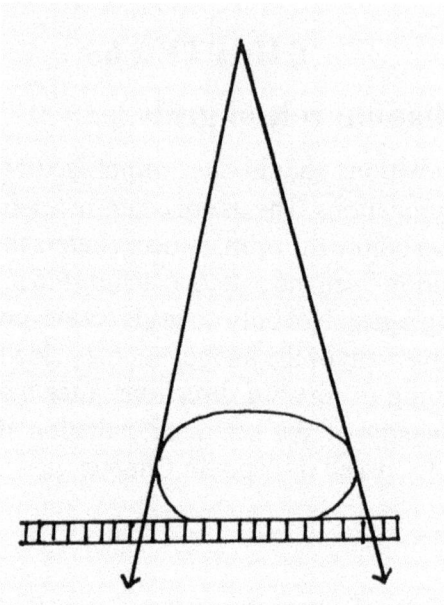

Figure 6a: Diagram to show spaces between lead-strips allowing the primary beam to pass through without being absorbed.

Grid ratio

A range of grids is available. Some are more efficient than others. The most important factor is the *ratio* of a grid; this is the relation between the depth/height of the lead strips and the width of the transparent spacers sandwiched between each strip [Figure 6b]. For example, if the height of the lead strip is 8 times the width of the inter-space, then the grid ratio is 8:1. If the height is 10 times the width of the inter-space, then the ratio is 10:1. The greater the grid ratio is, the more efficient the grid is in absorbing scattered radiation. However, some primary rays also get absorbed by a grid. Manufacturers clearly mark the ratio on at least one side of every grid and this is important for selecting the correct exposure factors as discussed in Chapter 7.

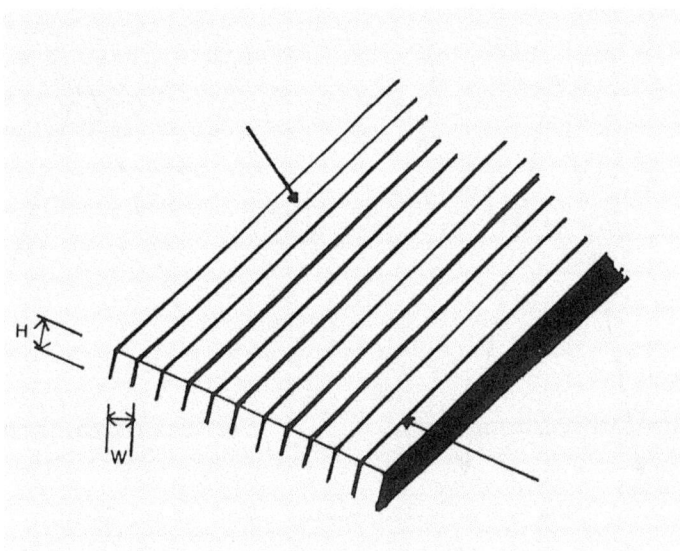

Figure 6b: Diagram showing width and height of grid strips. Arrow on left indicates inter-space. Arrow on right indicates lead-strip.

Parallel and focused grids

The most common types of grids for general radiography are "parallel" and "focused".

- Parallel grids consist of parallel lead strips [Figure 6c] resulting in absorption of some of the oblique rays of the primary beam as well as the scattered radiation

- Focused grids consists of lead strips that are progressively angled to accommodate the oblique rays of the primary beam. The angled strips converge to a point; the distance from the point to the grid is its focal distance, or length. The focal length of a grid is always clearly marked on it. Thus, a focal length of 100 cm means that the grid is effective in "cleaning up" scattered radiation without affecting the primary beam significantly at an FFD of 100 cm. If the FFD were shorter or longer then indicated as the focal length of the grid, then absorption of the primary rays would increase. This is called "cut-off" [Figures 6 d, e ,f].

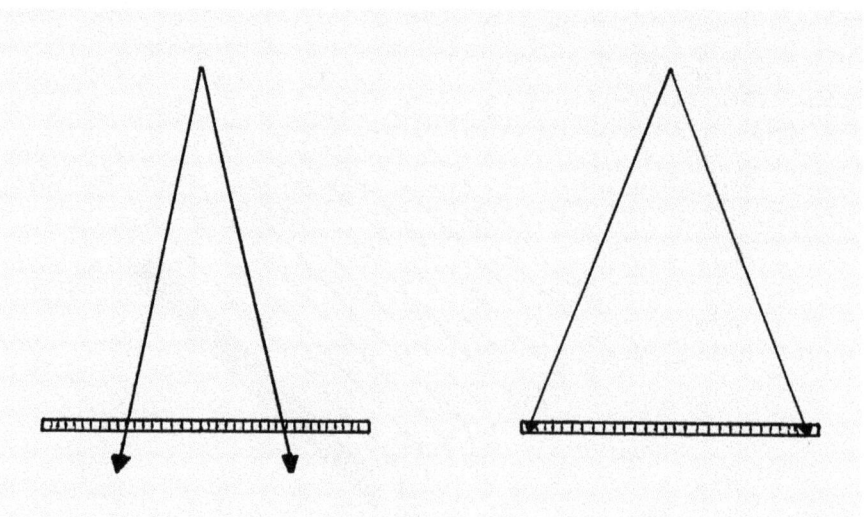

Figure 6c: Example of beam passing through spaces in a parallel grid. Note the absorption of outer oblique rays in right diagram.

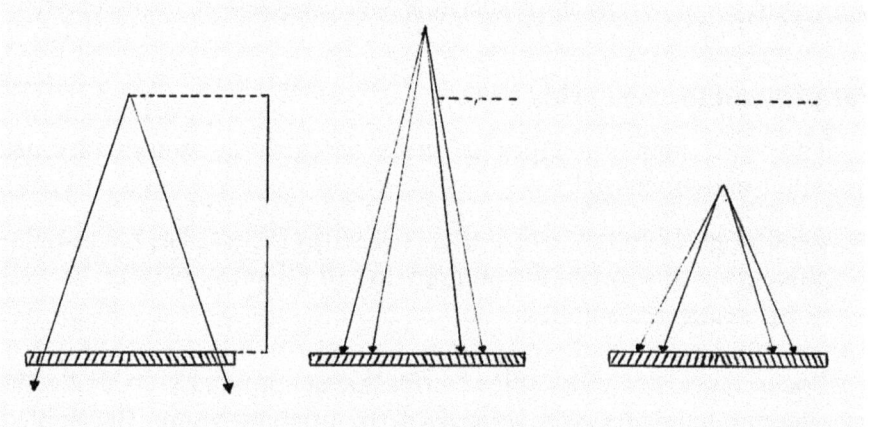

Figures 6 d, e, & f: (left) rays pass through inter-spaces of focused grid at correct focal distance. Increased distance (middle) shows some grid cut-off. Right diagram shows cut-off when distance decreased. To minimize retakes it is important to use correct FFD for focused grids to prevent grid cut-off.

Grid cut-off

Ideally, the central ray of the primary beam should coincide with the centre of the grid so that the primary rays intersect the grid perpendicularly. If this is not the case, a certain amount of "cut-off" occurs causing a progressive decrease in the transmitted X-ray intensity towards the outer edges of the grid [Figure 6 h].

Cut-off can result when the tube is tilted laterally across the lead strips [Figures 6 i, j]. This is a common problem when the cassette is not placed on a flat sturdy surface.

Parallel grids tend to produce cut-off when used at long distances from the X-ray tube. Care should be taken when placing a parallel grid to avoid cut-off.

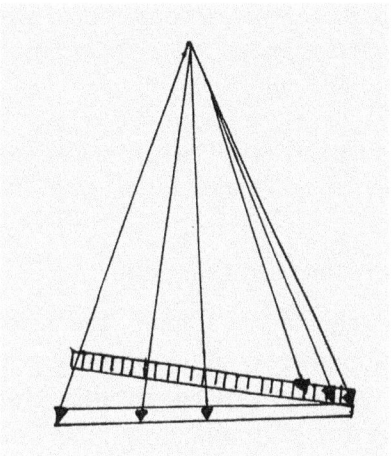

Figure 6h: Line diagram showing grid cut-off when central ray is not perpendicular to the grid. This problem occurs when using a focused grid in ward radiography. For example, cassette and grid placed under patient lying on a soft mattress results in non-alignment of central ray and lead-strips of grid.

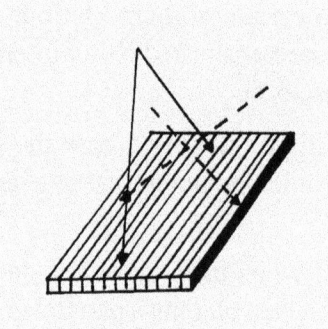 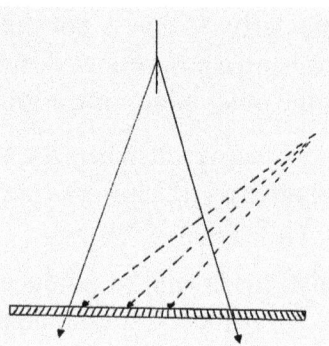

Figure 6i (left): Typical problem of tube tilted laterally across lead strips resulting in grid cut-off thus patient subjected to extra dose as radiograph has to be repeated without grid cut-off.

Figure 6j (right): Central ray tilted across lead strips causes cut-off. Note: When this occurs the primary beam is absorbed by the lead strips.

Focused grid: importance of tube-side

As indicated in the above figures, the lead strips are angled to accommodate the oblique rays of the primary beam to pass through the spacers and reach the film. To ensure that grids are correctly used, manufacturers mark the grid indicating tube side. If a focused grid is inadvertently placed upside down, most of the primary rays will be absorbed by the lead strips [Figures 6 k and l].

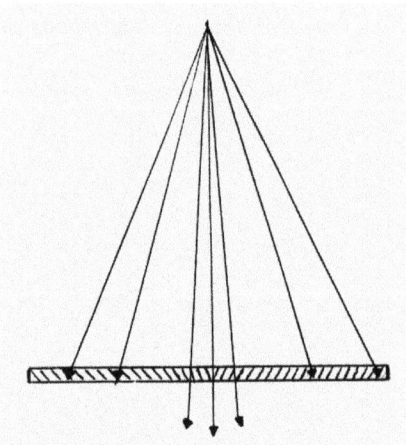

Figure 6 k: Line diagram showing absorption of primary beam because focused grid is incorrectly placed, i.e. upside down resulting in tube side of grid facing the film/receptor.

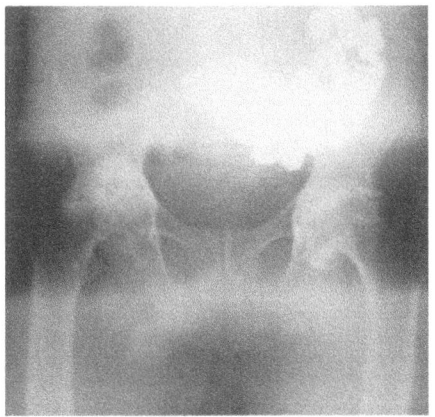

Figure 6 l: AP pelvis showing extensive grid cut-off because focused stationary grid was put up-side down by mistake causing insufficient implementation of ALARA as the radiograph had to be repeated.

Stationary grids

When performing out of bucky tray work, a stationary grid is used. The grid is placed on top of the cassette and positioned under the patient's body part being examined. When a stationary grid is used, grid lines may be evident on the resultant film [Figure 6 m]. The thinner the grid the less obvious the grid lines. Stationary grids can be used with cassette holder trays under the examination/couch table. For example, fine stationary grids may be supplied in skull units that do not include moving grid mechanisms.

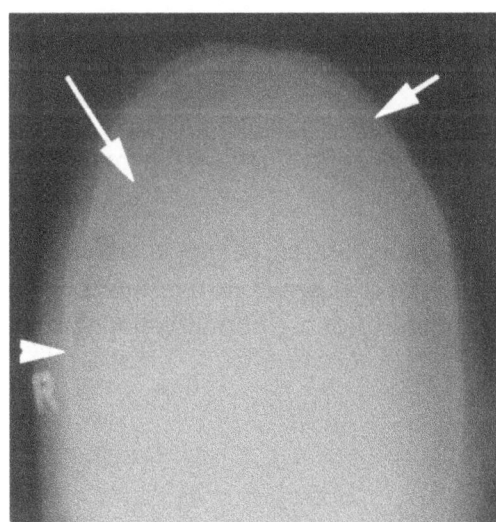

Figure 6 m: Arrows indicate stationary grid lines.

Moving grids

These grids move during the exposure to blur out the grid lines thereby improving visualization of image detail. Some grids move from one side to the other, whilst other move constantly from side to side and are called reciprocating mechanisms. When using moving grids, the exposure time must not be too short to avoid producing a striped pattern [Figure 6 n]. As with stationary grids, the "tube side" of a moving grid must face the tube [Figure 6 o].

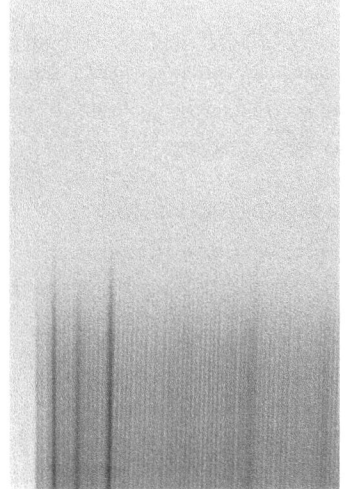
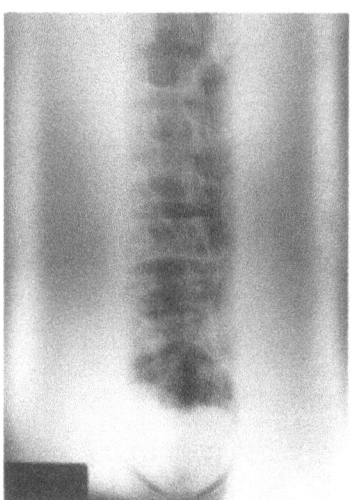

Figure 6 n (left): Too short exposure used together with a moving grid, results in thick black grid lines on a part of the image.

Figure 6 o (right): Radiograph AP abdomen showing only centre area of film density. Here, a focused moving grid had been placed upside down causing absorption of oblique primary rays.

Scatter clean-up: role of grids

As thick body parts produce more scattered radiation [Figures 6 p, q], grids are normally used to reduce this.

When examining (i) infants and small children, (ii) upper extremities of adults to mid region of humerus, and (iii) lower extremities to mid-thigh region of humerus on thin patients, grids are normally not needed, nor recommended.

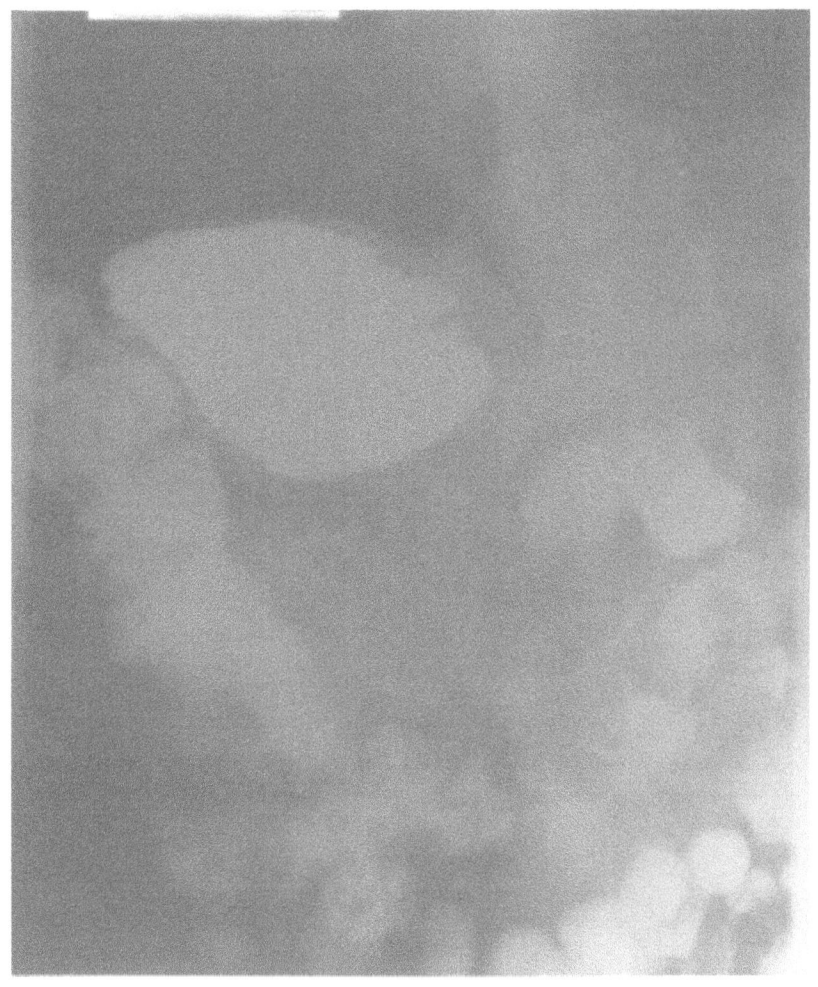

Figure 6 p: Fogged barium meal radiograph with very low subject contrast, i.e., differences in film densities. Image quality is poor because no grid was used; the technician forgot to replace it after a routine service. Scattered radiation caused 'fogging' of film.

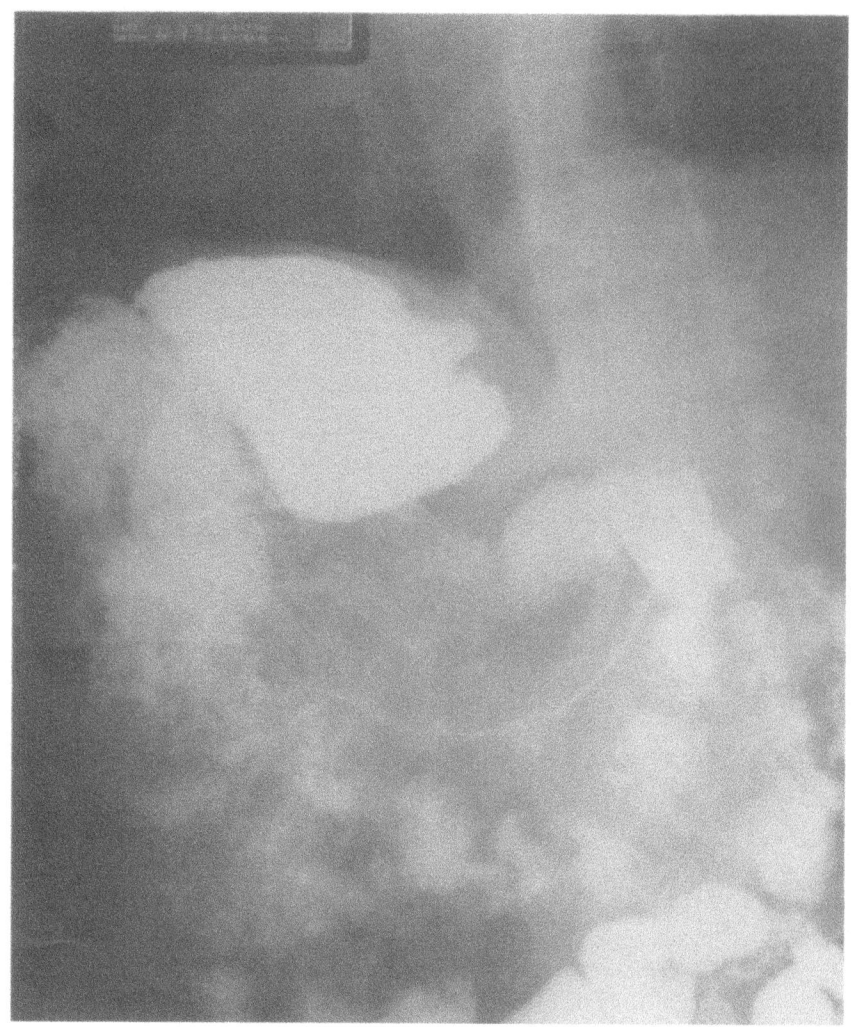

Figure 6q: Barium meal repeated after the grid was replaced. Note radiographic contrast greatly improved; scattered radiation absorbed by the grid.

Care and maintenance of grids

It is advisable to test a new grid for cracks before accepting it from the supplier. Also check that the grid performs according to its specifications when used at correct focal length. Thus, a 150 cm focused grid should not produce cut-off when used at this FFD.

Grids should be handled with care. When not in use, they should be stored in upright positions to avoid possible damage. Placing heavy objects on a grid could cause it to bend or crack. Keep sharp instruments away from grids as these could pierce the lead strips. Grids should not be dropped as this could result in cracks or bends resulting in non-alignment.

To prevent film artefacts, the outer casing of a grid should be cleaned using a damp cloth. Testing of grids should be included in the quality assurance programme of the department to ensure the grids function optimally. A bent grid results in grid cut-off and may require the patient to be unnecessarily re-exposed to ionizing radiation.

Quality assurance: grid tests

Grids should be checked at regular intervals to evaluate their efficiency. A poorly functioning grid does not benefit the patient.

Method

- Inspect the grid for bends or cracks and if found, take a radiograph of the grid;

- Place the grid on top of an unexposed, loaded cassette of the same size or slightly bigger than the grid. The grid must be placed with tube side facing the tube. Cassette must be on a flat surface so that the central ray of the X-ray beam will be perpendicular to the grid;

- Select the relevant FFD to match that of the focal length of the grid;

- Centre the tube to the centre of the grid. Central ray must not be tilted;

- Expose using a low exposure, e.g. exposure factors for adult finger or hand;

- Process the film and view the radiograph [Figures 6 r and s];

- Defective grids should not be used as they could contribute to production of poor image quality.

Figure 6 r (above) and Figure 6 s (below): Examples of malfunctioning grids. A radiograph of an efficient grid should show thin grid lines with overall even film density.

Grid factors

Since grids absorb scattered radiation contributing to the overall density of the film the exposure has to be adjusted. Some of the primary beam is also absorbed by the grid causing reduced density of the film. Consequently,

- the overall beam intensity (mAs) must be increased to compensate for loss of film density when a grid is used;

- the grid ratio determines how much the mAs has to be increased. Half the grid ratio, called the grid factor, is the amount by which the mAs must be increased;

- an 8:1 grid ratio has a grid factor of 4, whereas a grid ratio of 12:1 has a grid factor of 6. When using the latter, the mAs would have to be increased 6 times compared to non-grid exposure factors; a grid factor of 4 would require a 4 times increase in mAs.

In other words the more efficient a grid is in absorbing scattered radiation, the more the mAs has to be increased to compensate for loss of film blackening (i.e., density). This aspect of the grids is covered in Chapter 7.

Summary

When examining thick parts of the body, it is essential to minimize the scattered radiation for overall improvement of the image quality. Grids must be aligned correctly to the central ray to prevent grid cut-off. Grids should be checked regularly as a part of a quality assurance programme to reduce possible need to re-expose patients to ionizing radiation. When using grids, the mAs is increased when compared to non-grid exposure factors to compensate for loss of overall film density produced by both the primary beam and scattered radiation. Grids are costly items and should be handled with care.

CHAPTER 7

Radiographic technique, exposure factors, and quality assurance tests

A radiograph consists of a range of film densities called radiographic contrast. The most important factors influencing film density, or contrast are:

- The tube current, i.e. milliamperage (mA) discussed in Chapter 2 ;

- The distance between the tube and the patient and film, i.e. application of the inverse square law;

- The quality of the beam, i.e. kilovoltage (kV), and voltage waveform (output of generator) mentioned in Chapter 2.

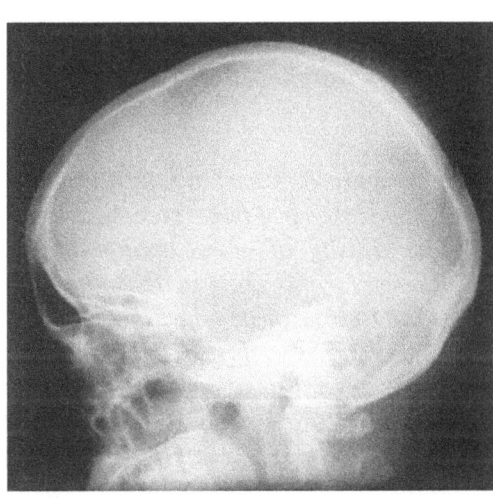

Figure 7a: Lateral skull radiograph with high contrast (short scale); insufficient kV used thus dense skull bones were not penetrated.

Due to the unique anatomical structures of each patient, the contrast of the resulting images is called subject contrast, and can be altered by manipulating various factors, such as exposure factors, use of beam limiting devices, use of compression, and introduction of a contrast medium. The reason for altering these factors is to comply with the ALARA principles.

An image with mostly white and black areas (i.e. large differences in film density) is defined to be of *short scale contrast*; [Figure 7a]. If an image has shades of grey, it will have a *long scale contrast* [Figures 7b and c]. These different shades of contrast of the image depend on the selection of kV and the output of the generator.

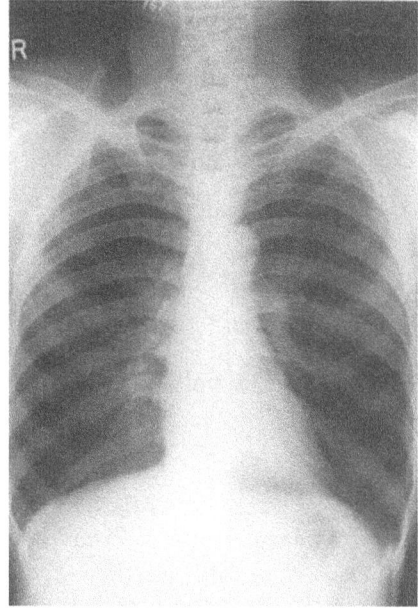

Figure 7 b: High contrast/short scale chest radiograph obtained using 46 kV and 25 mAs, no grid. Low kV results in high absorbed dose. Image does not have range of film densities.

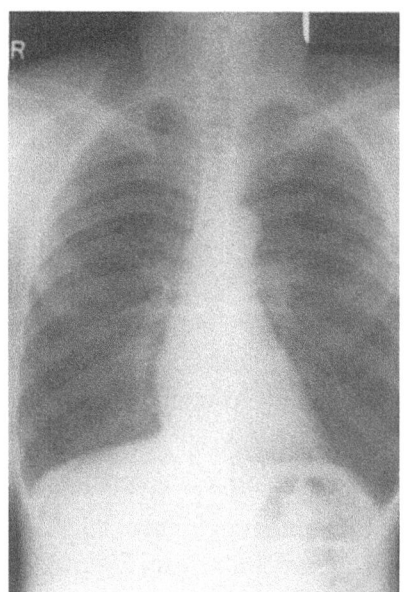

Figure 7c: Chest radiograph of same patient taken 6 months later. Note overall film density of more shades of grey. Reason for this short scale contrast film could be exposure factors changed to 85 kV and 2.5 mAs; dose to patient reduced due to less absorption of primary beam and use of low mAs. Conversion of factors based on rule of thumb: kV/mAs ratio, namely ↑10 kV then ÷ mAs by 2. A more scientific method: ↑ kV by 15% then ÷ mAs by 2.

The choice of required radiographic projections should be based on a correct assessment of the clinical indications and the expected yield from each examination and its contribution to further medical management of the patient. If a patient has recent previous films, these should be viewed before embarking on another examination.

Before commencement of any radiographic examination on a female patient of child bearing age, it is essential to determine whether she might be pregnant (see Chapter 8).

Patient positioning

To reduce the need for repeated examinations, care should be taken when positioning the patient. Immobilization devices, e.g. sandbags, compression bands, head clamps, should be used to reduce risk of patient movement during the exposure.

When examining the upper limbs of a patient, it is important to make sure that the patient's gonads are not within the primary beam [Figure 7d]. Beam restricting devices, as discussed in Chapter 5, should always be used to limit dose to the region of interest only. It is important to remember that the field size of the beam entering a patient produces a larger field size on the film because the beam diverges (fans out). So when considering the size of beam restriction (coning) to minimize dose, it is preferable to select size of required collimation before positioning a patient [Figure 7 e].

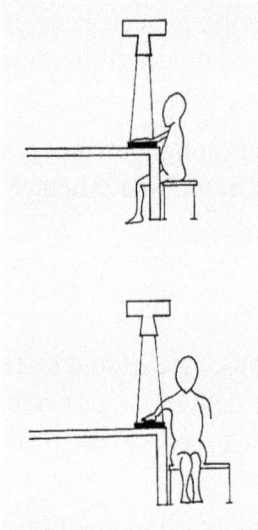 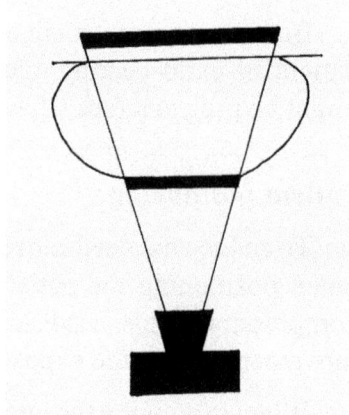

Figure 7d (left): Top diagram: Poor radiographic practices not in keeping with ALARA as the patient's gonads are in the primary beam. This example of incorrect patient positioning is not uncommon as some radiographers tend to forget basic radiation protection measures. Bottom diagram shows good implementation of ALARA.

Figure 7 e (right): Note field sizes. The one closest to tube is smaller than the one closest to film/image receptor.

Skull projections requiring caudal tilt of the tube, i.e. towards the feet of the patient, should be done with a very good restriction of the beam to minimize dose to thyroid. For example, a fronto-occipital 30 degree caudal tilt (Townes view) requires the patient to be facing the tube. If not well collimated, it may result in unnecessary radiation dose to the thyroid gland.

Skull radiography includes the orbits. Therefore care should be taken to reduce dose to eyes as mentioned in Chapter 2. Excessive radiation dose to the eye lenses could result in cataracts.

Chest radiography should whenever possible, be done PA to reduce dose especially to patient's eyes, thyroid, and breast tissues. Beam should be reduced to only expose the chest area (i.e. area of interest). It is essential to produce chest images taken on full inspiration to ensure that the entire lung fields are visualized [Figure 7 f].

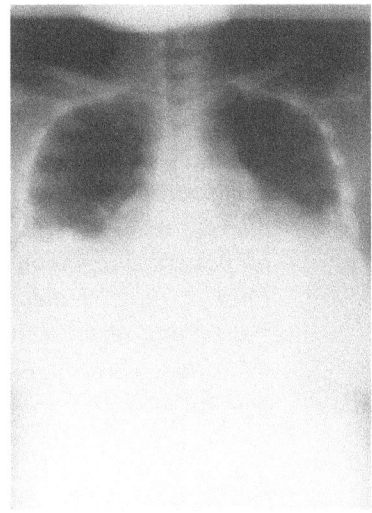

Figure 7 f: AP chest radiograph with minimal visualization of entire lung-fields because expiration film. Full inspiration chest radiograph is considered diagnostic for pattern recognition.

Most examinations require a minimum of two projections at right angles, such as AP and lateral. The patient must be correctly positioned to ensure the projections are at right angles. For example, when performing an AP radiograph of an ankle, it is important to check that the lateral and medial malleoli are equidistant to the film. This often requires slight medial rotation of the ankle and foot. Most patients find this position difficult to maintain. Thus, it is recommended to use sandbags to prevent the patient moving. Soft pads should be placed against the patient's ankle and supported by a sandbag.

Radiation protection includes the use of all methods to reduce dose to patient, staff, public and the environment. There are numerous publications on radiographic techniques pertaining to correct patient positioning, direction and angle of ray, and suggestions for use of immobilization devices [Figure 7 h]. One of them is the WHO Manual of Diagnostic Imaging: Radiographic Technique and Projections (ISBN 92-4-154608-5).

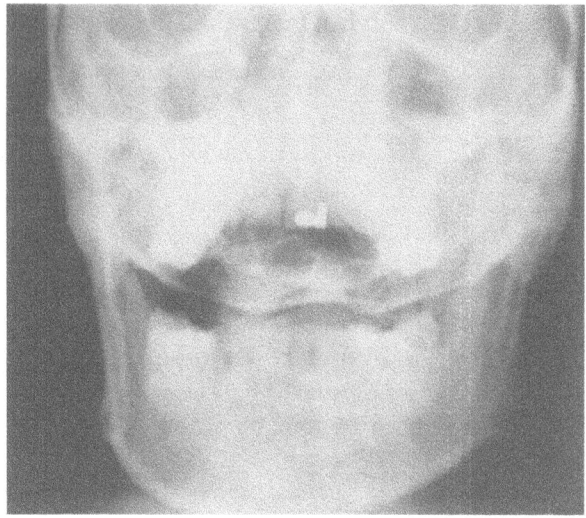

Figure 7 h: Inadequate open-mouth peg view: C1/C2 not visualized due to overlying dense skull bones. Film repeated to demonstrate area of interest, namely C1/C2 for pattern recognition. Poor patient positioning resulted in additional dose to eyes and thyroid.

Selection of kV

A change in kV results in a change in the penetrating power of the X-rays and the overall intensity of the beam. When kV is increased, shorter X-ray wavelengths with greater penetrating power are produced. Penetrating wavelengths have the ability to pass through dense structures, such as dense bones or dense plaster of Paris [Figure 7 i]. Because the beam penetrates dense structures, not much of the ionizing radiation is absorbed in the patient. Also, absorption of scattered radiation is reduced when the kV is increased [Figure 7 j]; the use of a penetrating beam is one method of reducing absorbed dose to a patient. However, this does not mean that all images should be obtained using high kV techniques as it may be necessary to visualize soft tissues and/or surrounding structures [Figure 7 k]. Soft tissue visualization requires use of low kV techniques [Figure 7 l].

Scattered radiation in high kV is energetic. Thus, the angle of the rays is not much different to those of the primary beam. Because high kV scattered radiation is not deflected very much, it means that it follows the path of the primary beam in a forward direction adding to film density. Use of relatively high grid ratios, e.g. 12:1 or higher if available, improves image quality as the scattered radiation in high kV techniques would be more efficiently absorbed.

Figures 7 m-n are examples of low kV *versus* higher kV imaging in radiography of knees; note there should be good detail of the patella on AP knee radiographs and this is usually achieved by adequate penetration of the primary beam [Figure 7 o].

Selection of mAs

The amount of mAs used to produce an image has a direct bearing on radiation dose to patient [Figures 7 p, q and r]. Use of a high mAs setting means increased dose to patient because the intensity of the beam is increased. The most important factors affecting mAs selections are:

- Focus film distance (FFD);

- Output of generator/capability of unit;

- Speed of film;

- Size, thickness, and type of phosphors in intensifying screens;

- Use of grid;

- Degree of collimation.

- Relationship to kV used;

- Size of focus.

Due to the inverse square law an increased FFD requires increased mAs to produce a comparable image based. For example:

- Increasing FFD two times, such as from 90 cm to 180 cm requires a four- times higher mAs.

Selection of mAs depends on output of generator to produce comparable images. When using a 3-phase generator instead of a single-phase one, a decrease in mAs is required. Thus, radiation dose to patient is reduced.

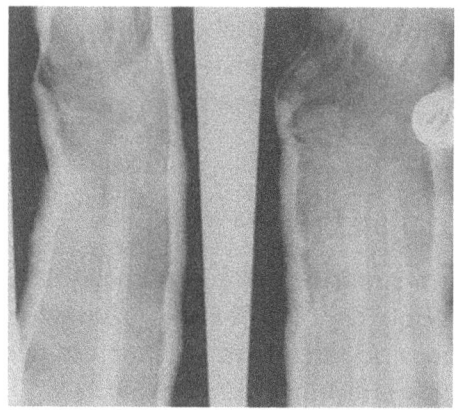

Figure 7 i: Inadequate penetration of wrist in plaster of Paris (POP). White/black film indicates that kV was not increased sufficiently to penetrate the POP. Patient received additional, unnecessary radiation dose as the examination had to be repeated.

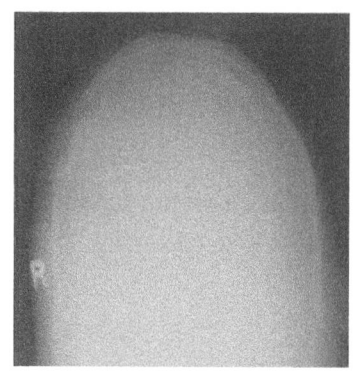

Figure 7 k: Low kV selected to demonstrate fracture of skull.

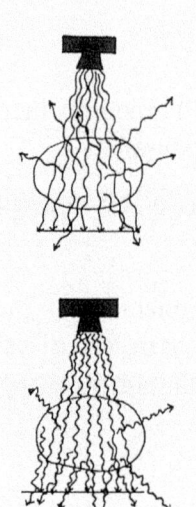

Figure 7 j: Low kV (top diagram) produces more scattered radiation than high kV (bottom diagram).

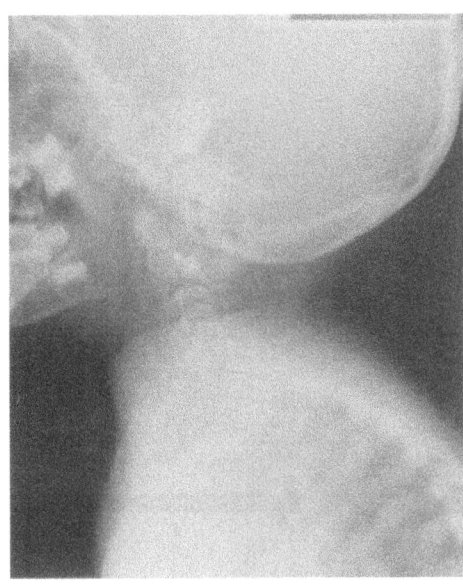

Figure 7 l: Low kV technique for soft tissue lateral neck examination on a child. Lack of evidence of beam restriction. Note artefacts (ear-rings). Retake done because not true lateral. Child received additional radiation dose to eyes and thyroid.

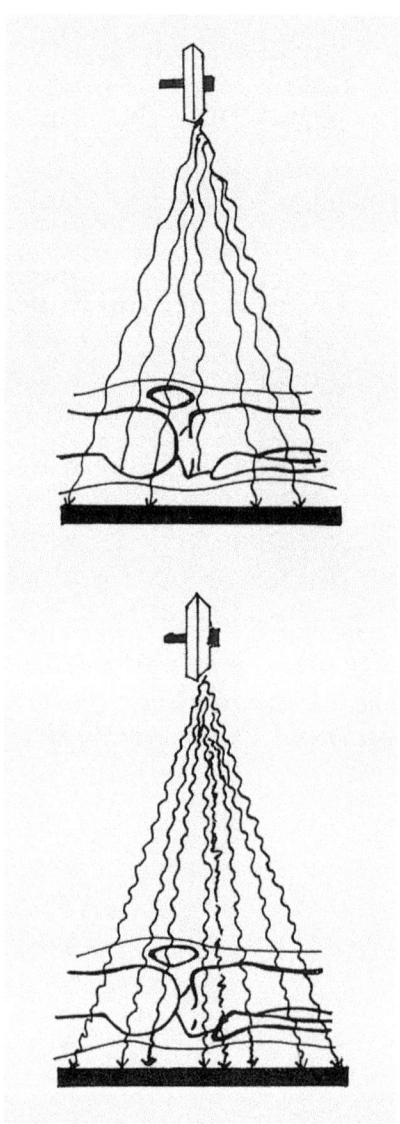

Figures 7 m: Line diagrams to show effect of low versus high kV: Low kV (top) results in high absorption of ionizing radiation whereas high kV (bottom) results in less absorption, or in other words, that more ionizing radiation reaches the film after having penetrated the tissue examined thereby causing a good visualization of the structures as required in pattern recognition.

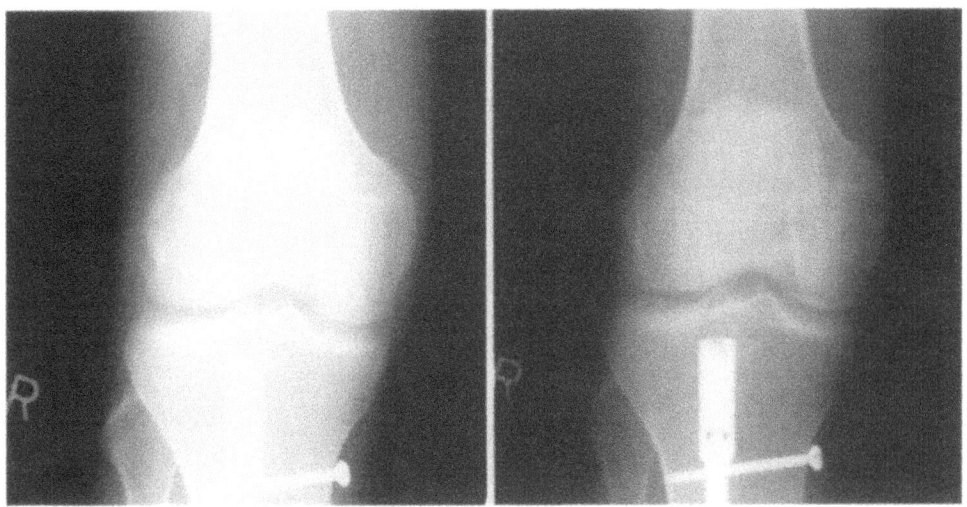

Figure 7 n: *AP knee (left) obtained using 55 kV and 9 mAs thus poor penetration of dense knee structures. Good penetration of knee structures (right) on follow-up images a few weeks later: exposure factors 70 kV and 3 mAs thus dose to patient reduced.*

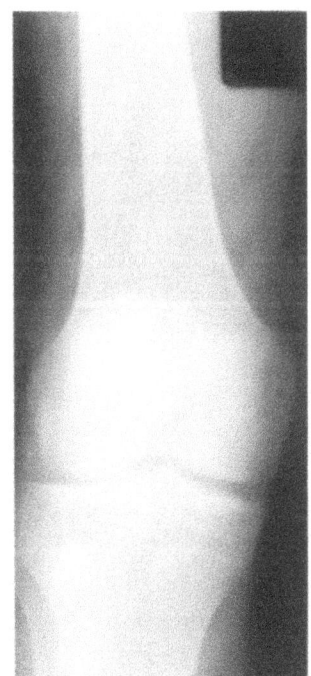

Figure 7 o: Poor patient positioning: entire knee architecture not included. Lack of penetration of knee joint due to low kV selection.

Figure 7 p: Very black film of distal femur due to high mAs. Patient subjected to unnecessary dose; film retaken with reduction of mAs.

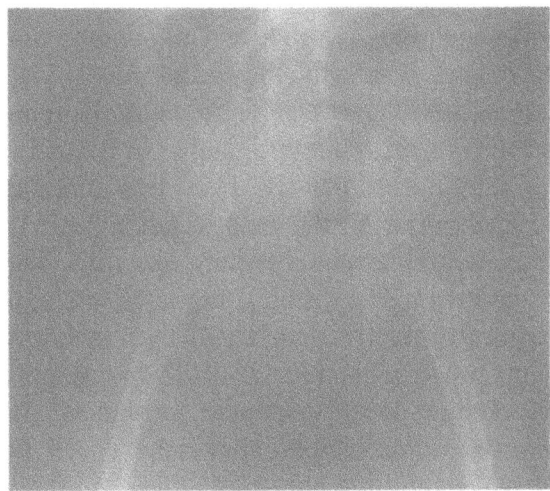

Figure 7 q: Implementation of ALARA not evident in black AP pelvis of child. Radiograph grossly over-exposed as mAs too high plus beam not restricted. Child's gonads in primary beam, which means dose to reproductive organs. Radiograph repeated.

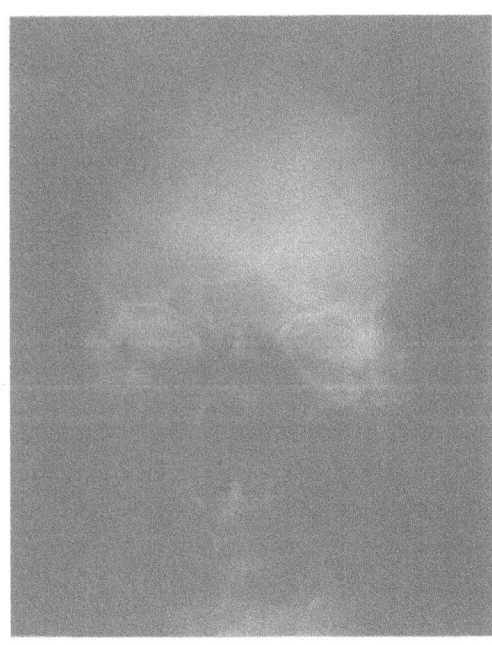

Figure 7 r: Black skull projection as mAs too high. Poor radiographic technique as patient's skull rotated. No evidence of beam restriction thus unnecessary dose to patient's eyes and thyroid. Poor implementation of ALARA.

Fast films need less mAs in order to produce comparable film blackening compared to slow films. The same applies to intensifying screens. Fast screens require less mAs slow ones. Thus, when changing from a 200 speed system to a 400 speed system original mAs needed is reduced by 50%, meaning that the patient receives less ionizing radiation. Selection of mAs is determined by the intensification factor of the intensifying screens. Thus, large phosphors require less mAs than smaller ones to obtain the same film blackening. Similarly, a single intensifying screen would require more mAs than having two screens in a cassette. Slow screen systems produce good image detail, but as more mAs is needed for film blackening, dose to patient is higher. The selection of specific film-screen combinations should be based on information required for pattern recognition purposes.

It would be of no benefit to the patient if a system is used that does not have good film-screen contact as information would be lost [Figure 7s]. Testing film-screen contact should be included in the department's QA programme to avoid subjecting patients to unnecessary ionizing radiation.

Figure 7 s: Arrows indicate poor film/screen contact.

When using a grid to minimize scattered radiation, the mAs must be increased based on the grid factor as discussed in Chapter 6. Limiting the beam to the area of interest requires an increase in mAs as discussed in Chapter 5. Figures 7 t and u clearly show how beam restriction without an increase in mAs reduces film density. Suggested increases in mAs when reducing film size:

- 35 x 43 cm field size to 25 x 30 cm: ↑mAs approximately 15%.

- 35 x 43 cm field size to 20 x 25 cm: ↑ mAs approximately 30%.

- 35 x 43 cm field size to 12 x 18 cm: ↑ mAs approximately 50%.

- 24 x 30 cm field size to 8 x 10 cm: ↑ mAs approximately 50%.

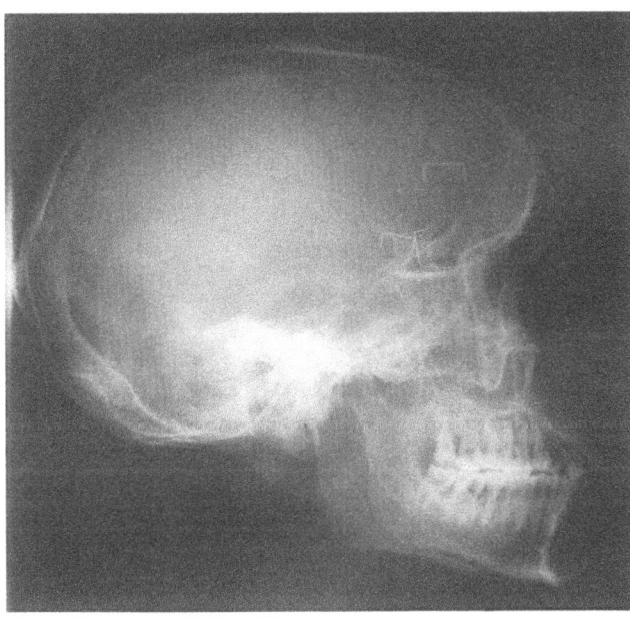

Figure 7 t: Lateral projection of skull phantom. Exposure factors: 66 kV, 24 mAs, grid, at 100 cm FFD. Adequate radiographic contrast obtained.

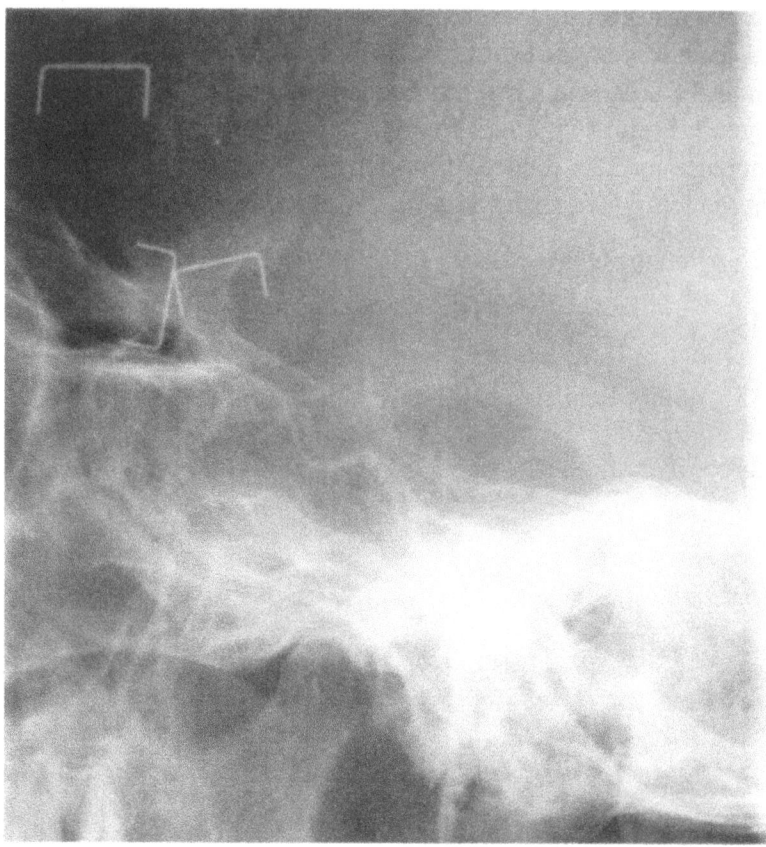

Figure 7 u: Coned lateral pituitary fossa of a skull phantom: extensive beam restriction without altering exposure factors were used for Figure 7 t. This shows that film density decreases when the film is reduced to reduce area of scattered radiation. It is necessary to increase mAs when restricting beam to compensate for loss of film density from scattered radiation.

Selection of high mAs (e.g. > 100) results in a great amount of heat production in the tube. Thus, it may be necessary to select broad focus when undertaking radiography on large patients as it may not be possible to use small focus due to tube limitations. Manufacturers include safety measures to prevent overheating of the tube although some units require one to check the tube rating and cooling charts applicable for each unit.

> **Suggested mAs adjustments when using different output generators:**
>
> **Low to higher output**
> - 2 pulse to 6 pulse, divide mAs by 1,5
> - 2 pulse to 12 pulse, divide mAs by 2
> - 6 pulse to 12 pulse, divide mAs by 1,2
>
> **High to lower output**
> - 12 pulse to 6 pulse, multiply mAs by 1,2
> - 12 pulse to 2 pulse, multiply mAs by 2
> - 6 pulse to 2 pulse, multiply mAs by 1,5

Exposure manipulation: kV/mAs

Each patient is unique in terms of size and shape and possible pathology. This requires exposure manipulation for optimal image production. There are two types of radiographic exposure charts that are used by most radiographers, namely variable kVp and fixed kVp charts, respectively.

Patient size needs to be ascertained when considering selection of kV to penetrate the anatomical structures. Measurements of a patient can be done by means of a calliper [Figure 7 v].

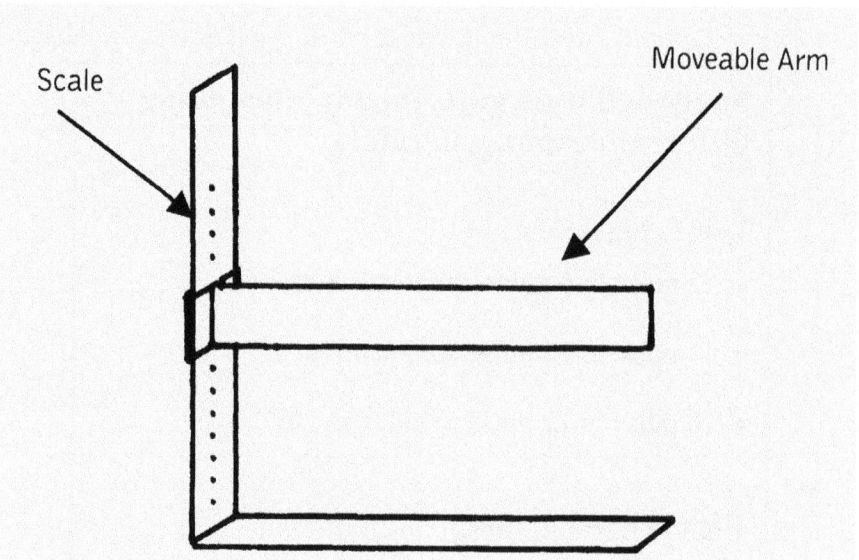

Figure 7 v: Drawing of a calliper used for calculating exposure factors based on patient's measurements as described by Lloyd in Quality assurance workbook for radiographers and radiological technologists (WHO/DIL/01.3).

Determining variable kVp charts

- Use the calliper to take measurements of patient along the line of the central ray;

- Chest measurement to be taken with patient seated or standing in a neutral position and normal respiration;

- If the part being measured is wedge-shaped, then take minimum and maximum measurements to obtain average by adding both measurements and dividing by 2.

Variable kV formula: Anatomical regions	Thickness X cm = Minimum kV
Adult bony tissue	Thickness cm X 2 + 27 = min kV
Adult chest	Thickness cm X 2 + 22 = min kV
Child bony tissue	Thickness cm X2 + 22 = min kV
Child chest	Thickness cm X 2 + 17 = min kV
Infant bony tissue	Thickness cm X 2 + 17 = min kV
Infant chest	Thickness cm X 2 + 12 = min kV
e.g. adult mid thigh measurement of 20 cm	20 cm X 2 = 40 + 27 = 67 kV

The above formula is a guide to establish minimum kV but this often means that the dose to the patient is high as the beam is not penetrating. For example, chest radiography should be done using high kV technique to reduce absorbed dose and for overall improved visualization of lung bases that are often not well seen if kV is too low [Figures 7 w and x].

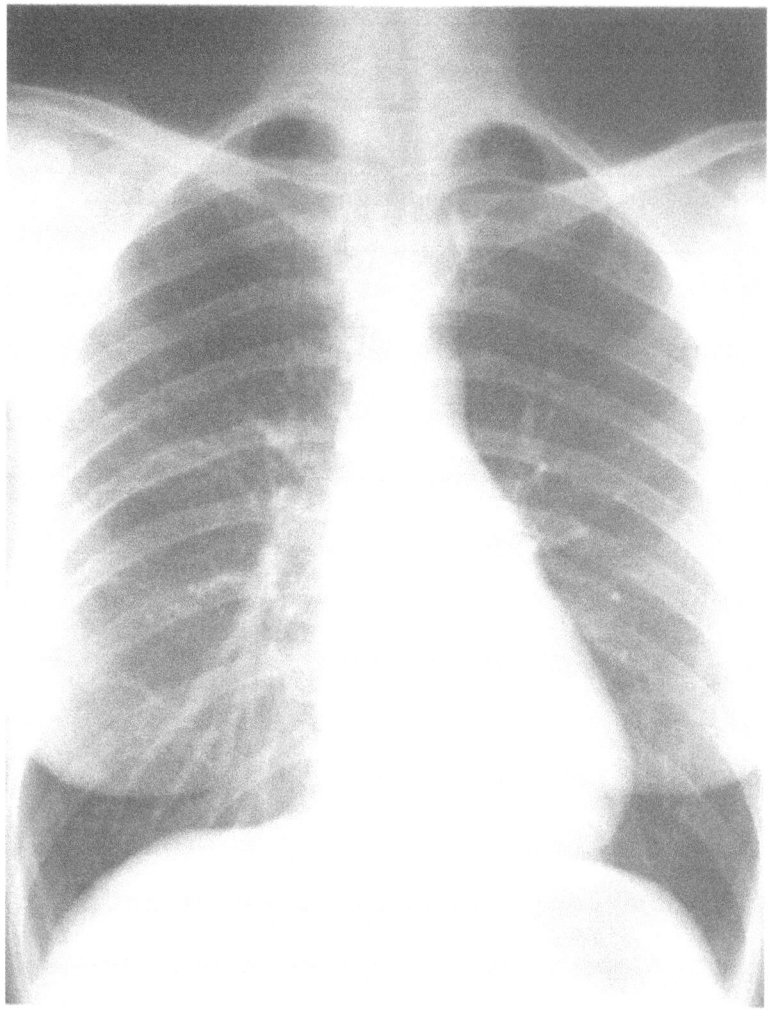

Figure 7 w: PA chest radiograph with exposure factors 70 kV, 16 mAs, 8:1 grid, at 180 cm FFD. Note: subject contrast is fairly high; lung bases not well penetrated. Compare this image with Figure 7 x of same patient.

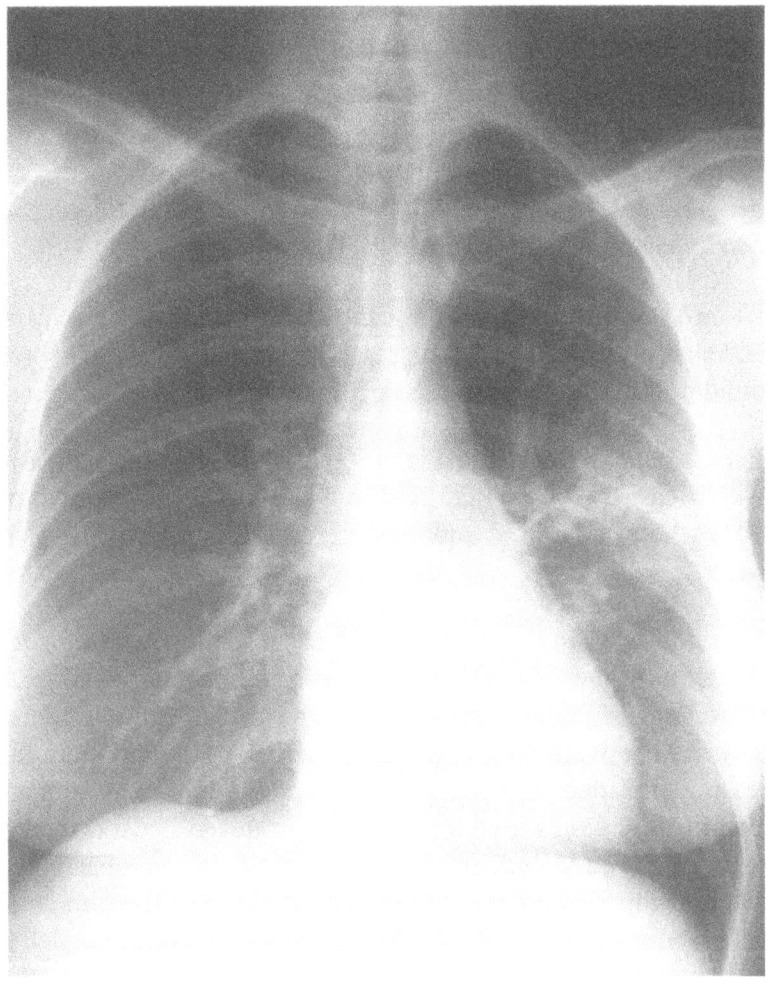

Figure 7 x: PA chest radiograph of same patient as in figure 7 w taken some years later. Note long scale contrast image obtained using 106 kV and 2 mAs. Overall improved penetration of bases with reduction of absorbed dose due to high kV and low mAs factors.

The variable kV technique requires adjustments, but it should be noted that the patient's clinical history needs to be taken into account to ensure adequate penetration. Hard to penetrate pathologies, such as sclerotic metastasis, osteoma, and radiation fibrosis require additional kV, whereas less kV is needed for more easily penetrated process, such as active osteomyelitis, degenerative arthritis, and osteoporosis

- **Fixed kV charts** are based on using a given kV then adjusting mAs to penetrate the structures. For example, adult shoulders could be examined using 65 kV, but if a patient is very obese then mAs would be increased. In this method, the dose to the patient could be high due to use of mAs for penetration of structures;

- **kV/mAs manipulation.** To use the highest possible kV may require adjustments to variable kV charts. An increase of 15% of kV requires dividing mAs by 2 to produce a comparable image;

- **Penetration of dense plaster of Paris (POP)** requires increasing the kV at least by 25% when POP is wet and freshly applied, and by 10% when POP is old and dry. Increase the mAs by 15% in both cases.

Deciding what to do in terms of exposure adjustments for repeats, often results in 'guess work' if a systematic approach is not adopted. It is therefore suggested that decisions for exposure adjustments should be based on what would be required in terms of either increasing or decreasing film density by a specific percentage. For example, if an original mAs of 4 was used and the film produced is dark, then a 25% decrease of the mAs will be 4 – 1 = 3 mAs. If a lumbar spine was examined using 80 mAs and there is a need to repeat the film due to the original being too dark, mAs should be decreased by 25% which would mean using 60 mAs. These adjustments are based on training and experience; it is recommended to always evaluate each image in terms of overall film contrast and film blackening by thinking of possible percentage changes to exposure factors.

If you are not sure what to adjust for the repeat, then the following 'rule of thumb' could be used: Increase kV by at least 15 and mAs by one third (33%) for lateral radiographs if the image quality of AP images is acceptable. For lateral projections of the skull, however, the above mentioned rule of thumb should be applied in a reversed way, i.e. the exposure factors used for the AP skull should be *decreased* for lateral projections by 15 and 33%, respectively;

- ***When in doubt,*** *increase kV instead of mAs to minimize dose to patient;*

Several factors may influence the mAs selection. Therefore, take one image and process it before taking additional radiographs. Guiding principle: keep mAs as low as possible to comply with ALARA.

Quality assurance tests to minimize unnecessary film fog

Various QA tests can be performed to minimize possible production of sub-optimal image quality due to film fog. Increased basic film fog may be due to poor safe-lighting, inadequate film storage conditions, and unacceptable processing conditions.

Safelight tests

It is important to ensure that darkroom safelight does not fog films. Unwanted film blackening (fog) causes reduction in radiographic contrast. Safelight tests should be done at least every six months to ensure they are properly working.

- Equipment for safelight tests:

An acceptable film screen light-tight cassette; black paper one-half the size of the cassette (2 sheets of black paper needed), clock (timer) with a second hand, box of unopened radiographic film, and general X-ray unit capable of selection of low mAs.

- **Step 1:** Cover lights on the processor and switch off all lights in the darkroom. In total darkness place an unexposed film in the light-tight cassette containing intensifying screens;

- **Step 2:** Expose the loaded cassette to radiation to obtain approximately the density reading of 1. Suggested exposure: half mAs of finger exposure but use same kV. Working tip: If it is not possible to select low mAs, then increase FFD using the inverse square law principle to determine the required mAs;

- **Step 3:** In total darkness open the cassette and remove the exposed film. Block off half of the exposed film using one sheet of black paper. This section of the film must remain covered throughout the test because this density is used as the control to ascertain whether the safelight is functioning correctly. Place the remaining sheet of paper on the other half of the film. Move the paper down to uncover part of the film to about midway. Switch on one safelight and expose the uncovered film portion for 60 seconds to the safelight as per normal working conditions and film handling in the darkroom. Remove this sheet of paper off the film and expose the remaining uncovered film half for further 60 seconds;

- **Step 4:** Remove the paper used to cover the other half of the film during the test. Process the film;

- **Step 5:** Place the film on an illuminator and inspect film density. This is a visual check to see whether there are differences in the densities of the film exposed only to radiation and the other half of the film exposed to radiation plus safelight. There should be very little visible evidence of increased film densities. The part of the film that was covered throughout the test, should have a density of approximately 1. Acceptable density limits when comparing the half side not exposed to the safelight and the half side exposed to the safelight, should not exceed 0.02 density for 60 seconds based on densitometer readings.

The above steps are to be repeated to check each safelight in each darkroom. Should unacceptable fogging be detected, then check (i) position of the safelight from the work bench (the height should not be less than 120 centimetres), (ii) safelight filter, and (iii) wattage of the safelight bulb.

The indicator light on the processor should be checked as per the above steps. The indicator light is to be uncovered and the safelight switched on during the tests. Density should not exceed 0.05 for a 2 minute exposure of film to indicator light plus safelight.

Processor control: performance monitoring

This quality control test should be done for all film processors to reduce unnecessary repeats. Monitoring film quality with regard to processing factors means assessing film contrast, film speed, and base fog as objectively as possible.

Equipment required

- sensitometer to expose film to different steps of light intensities;

- densitometer to measure optical density of selected sensitometric steps;

- thermometer to manually check the temperature of the chemicals;

- box of unexposed film;

- sheets of processing control charts or graph paper;

- **Step 1:** Under safelight conditions, expose one film to the sensitometer. It is important that blue light be used for monochromatic (blue-sensitive) film and green light for orthochromatic (green-sensitive) film. Process exposed film after checking temperature of chemicals as outlined in Step 2;

- **Step 2**: Temperature of chemicals to be checked using a manual thermometer. Temperature-gauge readings to be recorded. Note that both temperature readings should be the same; if there is a difference, then the faulty temperature gauge must be repaired as soon as possible to ensure the developing temperature is constant for optimal film processing;

- **Step 3**: Process the exposed film;

- **Step 4**: Using a densitometer, read densities for each sensitometric step on the film;

- **Step 5**: The sensitometric step with the density closest to 1.20 (mid-density), is to be used to determine speed index. For example, if sensitometric step 9 has density closest to 1,20 (including base fog), then all subsequent readings for speed index should be at this value (i.e. step 9);

- **Step 6:** To obtain contrast index, refer to the recorded density readings of all the steps. Since contrast is the difference between two densities, the contrast index is obtained by subtracting the density reading of one step from the readings of another step, and use is made of the density reading for the speed index, i.e. if step 9 is closest to mid-density, then subtract density reading of step 11 from step 9 to get contrast index. Same steps to be used for all future readings;

- **Step 7**: On graph paper, record temperature, date, and base fog reading;

- **Step 8**: Plot speed index and contrast index on graph paper;

- **Step 9**: For five consecutive days, repeat the preceding steps to obtain average density for the speed index and the contrast index. These become the controls against which all future sensitometric films will be compared. Plot average speed index and contrast index obtained over the five days on graph paper. An acceptable variation is **plus or minus 0.15** (on the special graph paper). Any deviation outside these lines means one or more processing factors are not performing correctly. For example, check replenishment-rates of developer, check the temperature as an increase/decrease causes changes to film density, check that chemicals have been correctly mixed in terms of specified quantities and that tarter has been added. If there is marked increase in base fog, then perform safelight test.

Careful film-handling and film storage

Exposed film is sensitive to light and care should be taken when handling film at all times.

Film should always be handled with clean hands and in a dust free environment. Film boxes must be placed vertically (i.e., "standing") in a cool room with good air circulation.

Boxes of film must never to be stacked flat on top of each other as this will cause marks on the films. Pressure, even very slight, on film causes 'white streaky' marks that could be incorrectly interpreted as pathology.

Summary

To reduce unnecessary radiation dose to patients, optimal image quality should be obtained using the highest possible kVp for visualization of the anatomical parts to be examined. If the examination requires visualization of soft tissues, low kV technique should be used. Patient dose is decreased when overall penetrating abilities of the X-ray beam are high (high kV). Patient dose is increased when excessive mAs selections are used. A radiograph which is very black, indicates that the patient received unnecessary radiation due to high mAs factors.

To minimize the need for repeats, it is recommended that basic quality assurance tests be undertaken. Film fog contributes to poor image formation and may result in non-visualization of detail. Illuminators (viewing boxes) should be clean and, if possible, each of them should have the same amount and type of illumination.

CHAPTER 8

Exposure to ionizing radiation during pregnancy

Patients and medical staff

Possible risks to the embryo and foetus must be considered when using ionizing radiation to investigate female patients of reproductive age. Alternative imaging modalities and techniques not involving ionizing radiation should also be considered.

A female patient of reproductive age who presents for an examination in which the pelvic area will be irradiated should be asked whether she is, or might be pregnant. If she is not sure, then it is advisable, except in an emergency, to reschedule the examination until a pregnancy has been excluded.

Pregnant radiation workers

Occupational exposure of pregnant radiation workers (radiographers/technologists) must be as low as possible, and complying with national laws and international recommendations. Most employers accommodate pregnant radiographers by allowing them to work in low risk workstations. The ICRP recommendations state that pregnant diagnostic radiographers should not be involved in examinations using mobile X-ray units, radiographic procedures in operating theatres, and fluoroscopic procedures. Pregnant radiographers should be provided with radiation monitoring devices in addition to the personnel monitoring devices, such as TLDs (thermo luminescent dosimetry).

Summary

Whenever possible, X-ray examinations of a pregnant woman should be postponed until after delivery. If the examination has to be carried out immediately, the radiation dose must be kept at a minimum, without comprising treatment of the patient. Further information can be found in the joint WHO/ICRP publication (WHO/DIL/02.1): "Ionizing Radiation for Diagnostic Imaging or Treatment during Pregnancy: Some practical advice".

Pregnant radiographers should be provided with additional radiation monitoring devices, and restricted to work in low risk areas as per the recommendations of the ICRP.

CHAPTER 9

Self-evaluation of images: application of ALARA

Radiation protection to reduce dose to staff, patients, and members of the public is achieved by legislation and education. The responsibility of implementing legislation resides with the national authorities. However, operators of ionizing radiation units are responsible for keep their work in line with the ALARA principles.

It has been stated that "good radiological practice is something that can be taught". This point of view is re-iterated by Rehani, namely that nearly 40% dose reduction is achievable by appropriate training. Hopefully this book provides some additional knowledge on this. This chapter presents a series of images to be used for assessing image quality, and to see how the ALARA principles can be applied.

Practical hints for self-evaluation of image quality and implementation of radiation protection measurements

Carefully assess Figures 9a to 9i and read the accompanying text for each figure to test your knowledge of factors affecting good radiographic practice, such as FFD, beam restrictions, and gonad protection. Turn to the end of the chapter for suggested answers.

- Figure 9a: Lateral thoracic spine.

- Figure 9 b: Lateral skull.

- Figure 9 c: Lateral skull.

- Figure 9d: AP abdomen.

- Figure 9 e: AP abdomen.

- Figure 9 f: AE (after evacuation) film.

- Figure 9 g: Delayed AE film.

- Figure 9 h: Lateral abdomen.

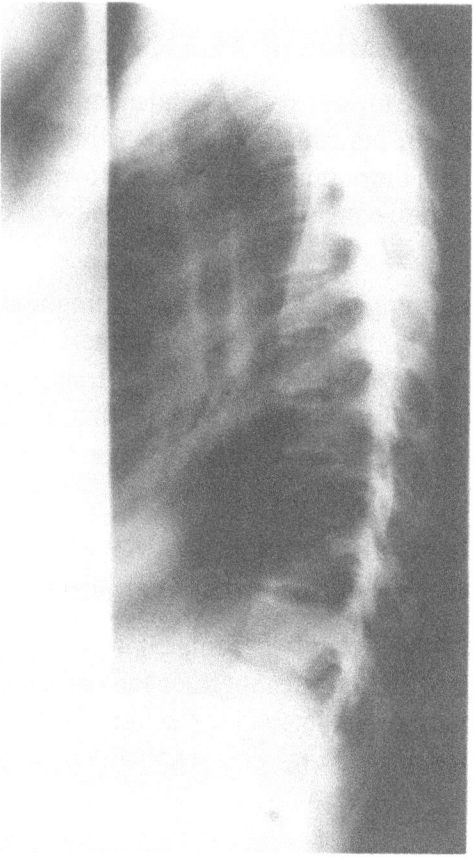

Figure 9a (left): Lateral thoracic spine. Is this a good image? Is there evidence of implementation of ALARA? Explain why the ribs are blurred. Why is there a lack of penetration and visualization of the upper lumbar vertebrae?.

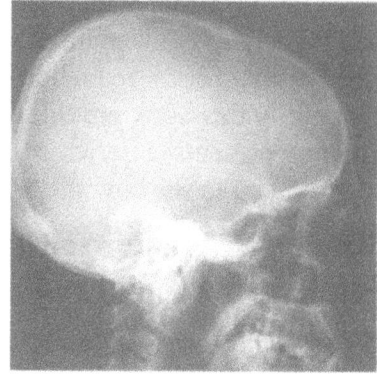

Figure 9 b (above): Comment on this lateral skull in terms of contrast and image detail. Would you repeat this radiograph?

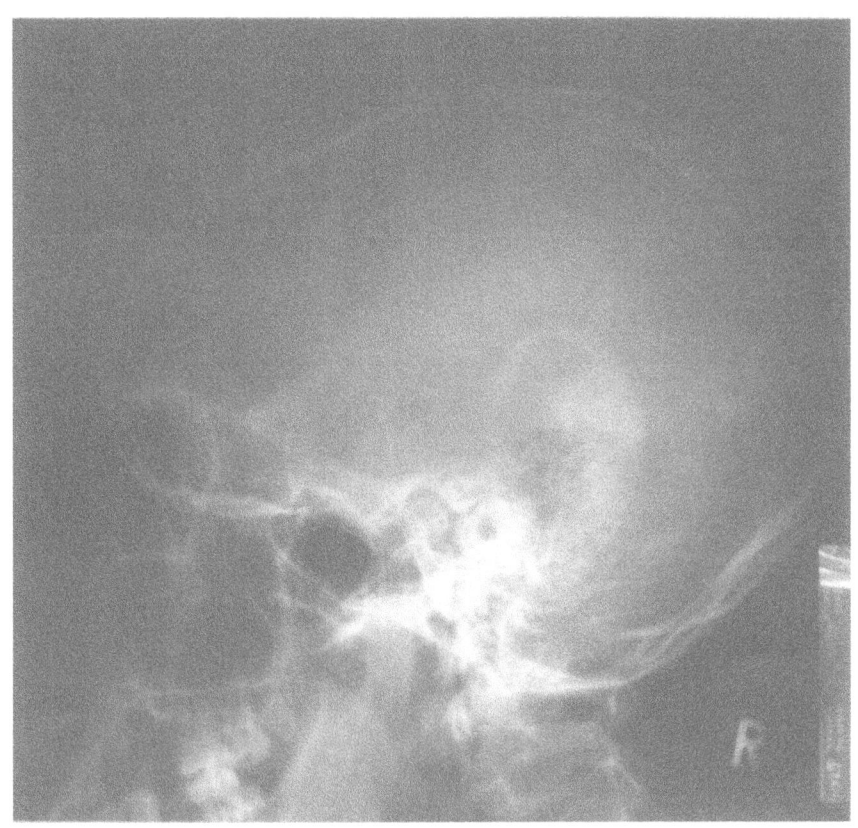

figure 9 c: Would you repeat this lateral skull radiograph? Consider ALARA and benefits to patient.

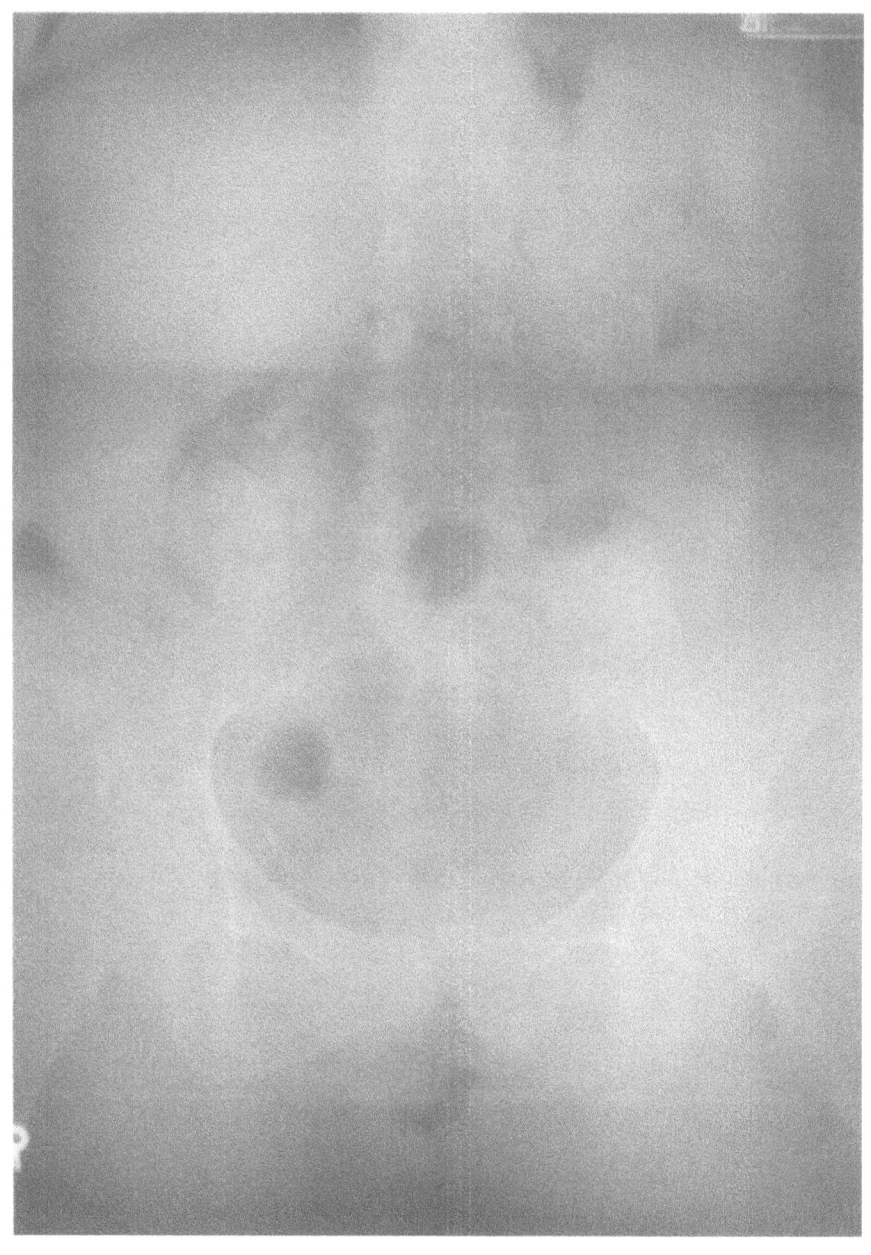

Figure 9d: AP abdomen. Which factors would you alter to achieve improved penetration of the abdomen? Would your suggested factors increase or decrease patient's absorbed dose?

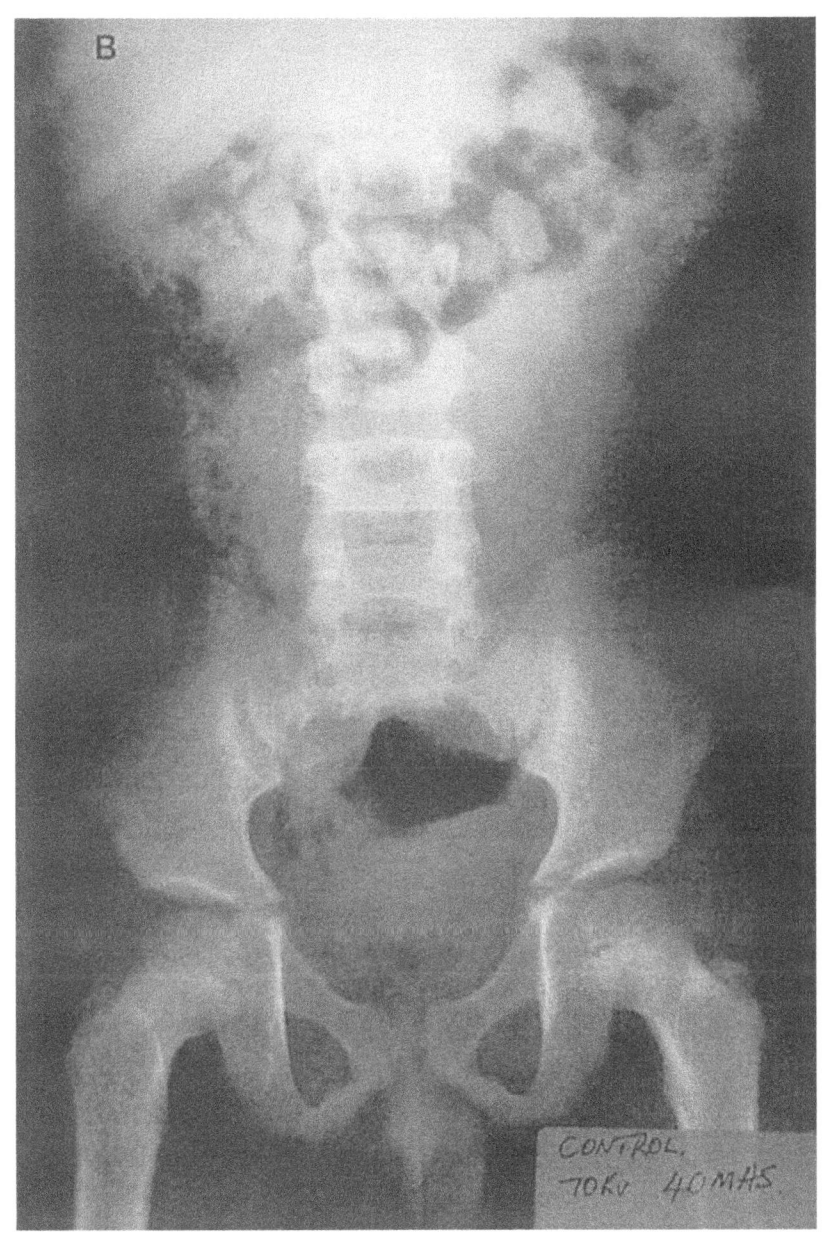

Figure 9 e: AP control abdomen of 10 year old male for barium enema. Is this image in keeping with ALARA?

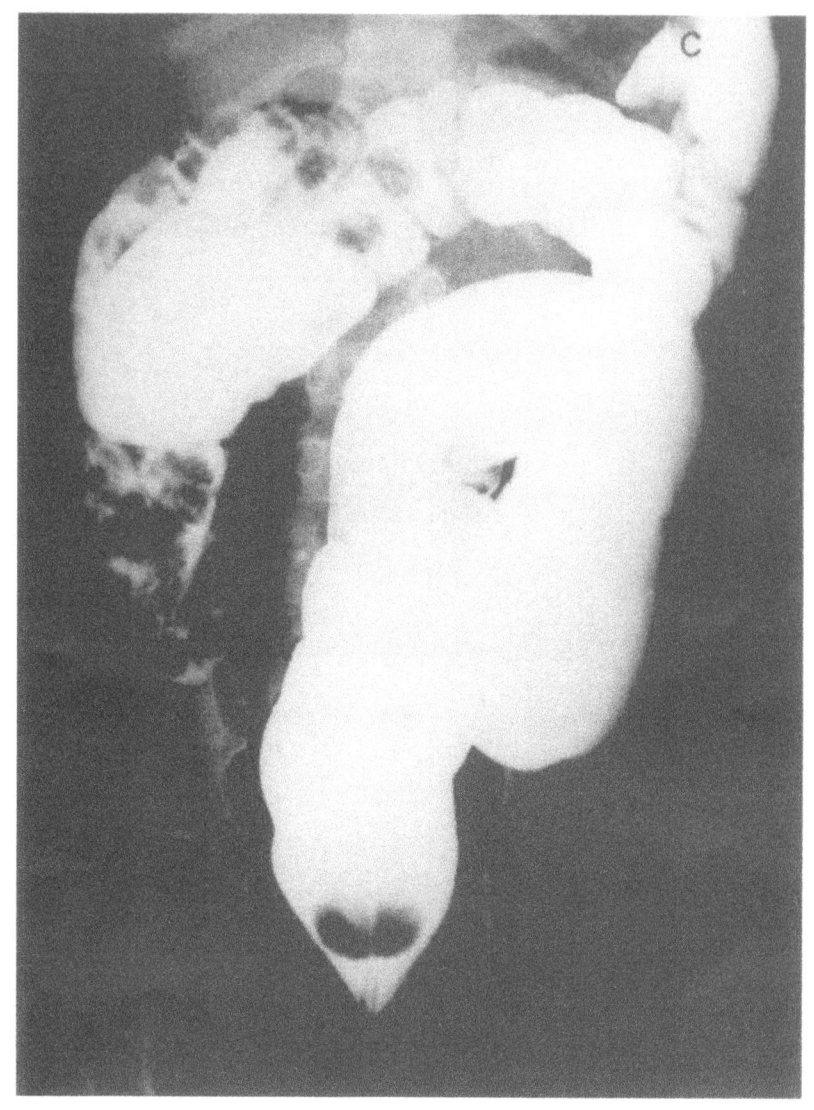

Figure 9 f: AE (after evacuation) film of same patient. Is this radiograph acceptable based on the ALARA principle?

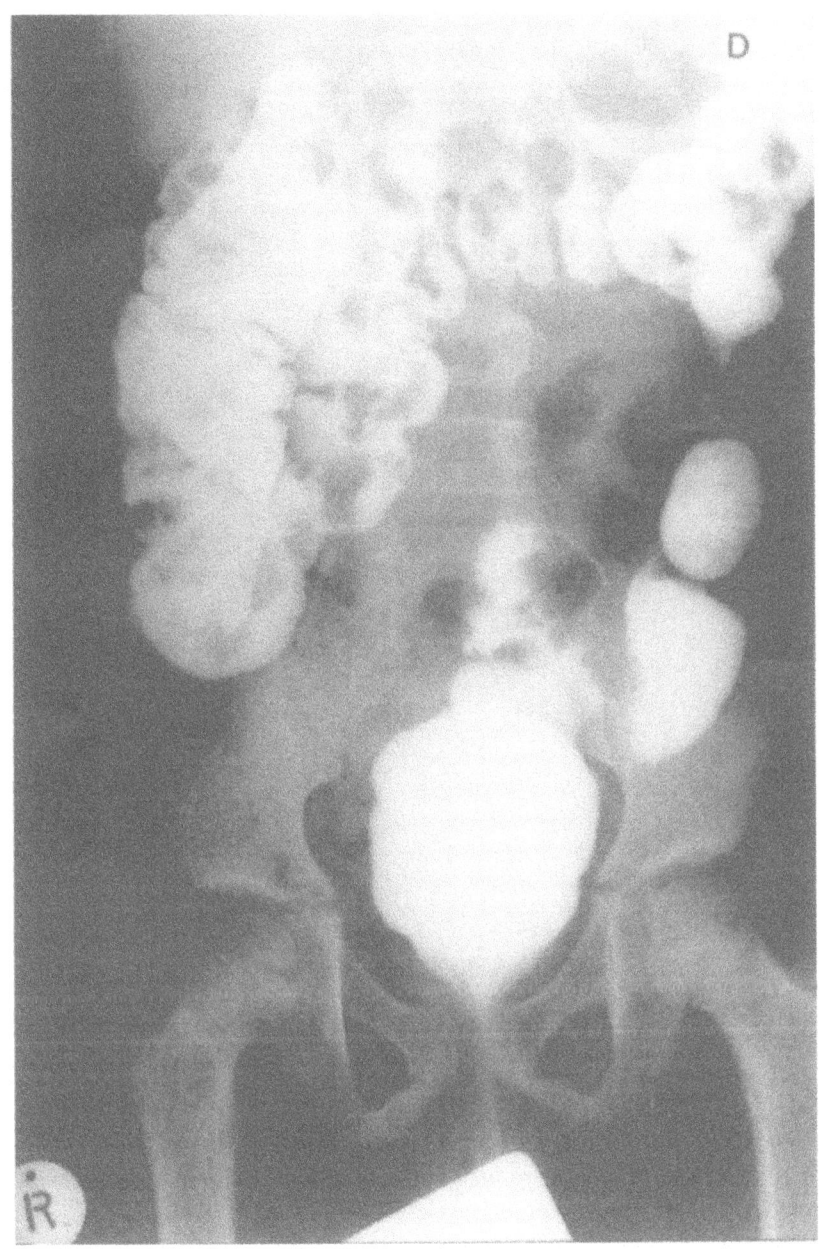

Figure 9 g: Delayed AE film of same patient. Note evidence of gonad protection on this long scale contrast image. Is this image in keeping with ALARA?

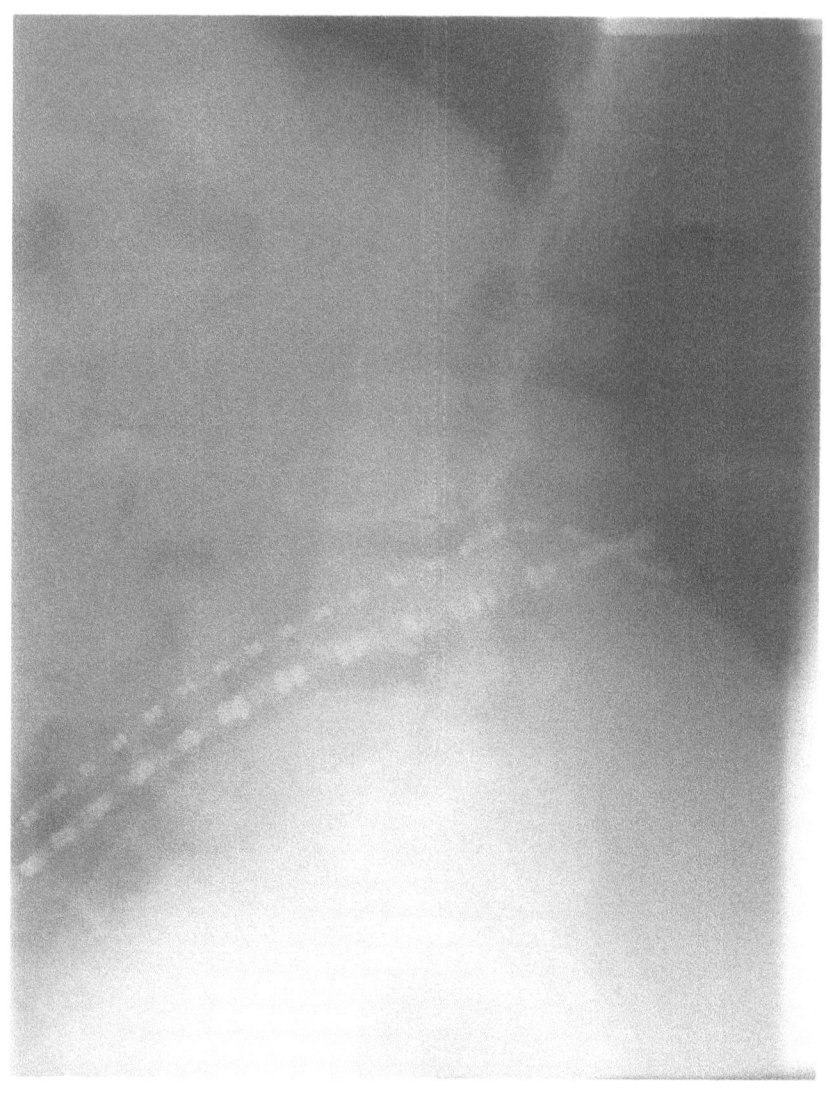

Figure 9 h: Note the detail of some beads is blurred. Why? Do you position a patient in a specific way to produce blurring of some anatomical structures in keeping with ALARA for good image detail and /or dose reduction?

Suggested answers to questions for Figures 9a to 9h.

- Figure 9a: Image acceptable. ALARA evident as beam restricted to area of interest. Ribs blurred diffusion technique (long exposure time with patient breathing) used to blur out ribs that obscure vertebrae. Lumbar vertebrae not penetrated because they are below the air-filled lungs.

- Figure 9b: Radiographic contrast and detail good. Not necessary to repeat the radiograph even though there is no evidence of beam restriction.

- Figure 9c: This radiograph should be repeated as patient was rotated during the exposure and overall film density is increased due to high mAs. Beam needs to be collimated.

- Figure 9d: Short scale image as not sufficient kV used. Radiograph is under-penetrated meaning that kV should be increased by at least 20% for optimal visualization of abdominal structures. Increasing the kV will result in a more penetrating beam with less radiation dose to the patient. Compression could be consisectiondered to reduce overall thickness of tissue thus reducing scattered radiation.

- Figure 9e: ALARA not applied as gonads of the young male not protected. Contrast short scale due to low kV and high mAs exposure factors which meaning dose to patient not limited.

- Figure 9 f: Gonads not protected and high kV/low mAs factors to reduce dose not used.

- Figure 9 g: Application of ALARA principles is evident; gonads protected and high kV technique used as long scale contrast image with overall improved visualization of bowel patterns.

- Figure 9 h: String of beads around patient's waist not removed; in the lateral position the beads closest to the film/tabletop well seen. The beads closest to tube, due to increased subject film distance, are blurred. Examples of using this principle of geometric un-sharpness include PA mandibles, and PA sterno-clavicular joints.

www.ingramcontent.com/pod-product-compliance
Ingram Content Group UK Ltd.
Pitfield, Milton Keynes, MK11 3LW, UK
UKHW051128200426
11947UKWH00040B/1551